Polynuclear Aromatic Hydrocarbons

NOMENCLATURE GUIDE

First Edition

K. Loening
Topterm
Columbus, Ohio

J. Merritt
Chemical Abstracts Service
Columbus, Ohio

D. Later
Mountain States Analytical, Inc.
Salt Lake City, Utah

W. Wright
Battelle
Richland, Washington

Published in cooperation with the International Polynuclear Aromatic Hydrocarbons Symposia

BATTELLE PRESS
Columbus • Richland

Library of Congress Cataloging-in-Publication Data

Loening, K.
Polynuclear aromatic hydrocarbons nomenclature guide / K. Loening...[et al.].

p. cm.
"Published in cooperation with the International Polynuclear Hydrocarbons Symposia."
Includes index.
ISBN 0-935470-59-X
1. Polycyclic aromatic hydrocarbons--Nomenclature. I. Loening, Kurt, 1924- . II. International Polynuclear Hydrocarbons Symposium.
QD341.H9P75 1990 90-39042
547'.61--dc20 CIP

Printed in the United States of America.

Battelle Press
505 King Avenue
Columbus, Ohio 43201-2693
614-424-6393
1-800-451-3543

CONTENTS

CONTENTS (Continued)

Preface

Research relating to Polynuclear Aromatic Hydrocarbons (PAH) is a multidisciplinary activity frequently carried out by scientists not familiar with the intricacies of chemical nomenclature. This became evident when the organizers of the Seventh International Symposium on Polynuclear Aromatic Hydrocarbons in 1982 asked me to conduct a workshop on the chemical nomenclature of PAH with a problem solving session on naming difficult structures. Since the program was already set, it was decided to make it a breakfast session. It was scheduled from 7am to 9am Wednesday morning, and some 30 people were motivated to arise before daylight to attend. The program was so successful that it quickly became a tradition.

This nomenclature guide is a compilation of the presentations given at these breakfasts. It concentrates on the PAH themselves, their nitrogen, oxygen and sulfur analogs including functional derivatives, the metabolic products of PAH, and enzymes. For each topic, references are provided to the original nomenclature recommendations to enable the reader to check out further details.

In this way by acquainting the readership with the recommendations of the International Union of Pure and Applied Chemistry (IUPAC) and the International Union of Biochemistry (IUB), it is hoped that this nomenclature guide will contribute to good communication between all the different types of scientists and the general public concerned with PAH.

The readership is encouraged to submit to me additional topics relating to PAH where nomenclature is a problem.

Finally I want to acknowledge my deep appreciation to my coauthors who helped make these breakfasts so successful.

Kurt L. Loening
Topterm
North American Division
P.O. Box 21213
Columbus, Ohio 43221

Polynuclear Aromatic Hydrocarbons

NOMENCLATURE GUIDE

First Edition

Chapter 1

SOME AIDS FOR NAMING POLYCYCLIC AROMATIC HYDROCARBONS AND THEIR HETEROCYCLIC ANALOGS

KURT L. LOENING AND JOY E. MERRITT

Interpreted in a very broad sense, the class of polycyclic aromatic hydrocarbons includes any hydrocarbon containing more than one aromatic ring and does not exclude hydrocarbon ring assemblies, such as biphenyl, bridged systems or spiro systems. However, the compounds of major interest to cancer researchers are generally what is called by the International Union of Pure and Applied Chemistry (IUPAC) *ortho* or *ortho-* and *peri-* fused ring systems (1a). Thus, the present discussion is limited to hydrocarbons and heterocyclic analogs of this subclass of polycyclic aromatic compounds.

The nomenclature of *ortho* and *peri*-fused hydrocarbons has been previously reviewed by Jacob and Grimmer (2). Other reviews, including the nomenclature of heterocyclic analogs, are those by Acheson (3), McNaught (4), and one written in Japanese by Hata (5).

The total number of possible constitutional geometric isomers of *ortho*- and *peri*-fused hydrocarbons containing up to eight fused benzene rings (polyhexes) has been calculated to be 1,448 (6); and it is obvious that by introducing other sizes of rings and heteroatoms the possibilities increase dramatically. Thus, the class of *ortho* and *peri*-fused ring systems is very large.

There is no single comprehensive, totally systematic, method currently in use for naming this class of fused ring systems. The most comprehensive system in use is that codified by IUPAC in "The Nomenclature of Organic Compounds" (1), also referred to as "The Blue Book". It is based on the early proposals of Patterson (7,8) and it is the system currently used by Chemical Abstracts Service (CAS) in its indexes (9). However, other than for the prefixes of the simple monocyclic hydrocarbons (e.g., cyclopenta, cyclohepta) the IUPAC system is a system built on trivial and semisystematic names. The attempt of Balaban (10,11) for total systematization has so far only encompassed benzenoid types. The nodal nomenclature system being developed by Lozac'h *et al.* (12) seeks to encompass all acyclic and cyclic systems in a single systematic nomenclature, but at present it is limited to the naming of graphs without specification of the actual atoms involved (nodes).

Limitations of the earliest proposed localized systems of Bally and Scholl (13), and Stelzner and Kuh (14) are noted in the reviews cited above and by Clar (15) who also proposed another limited semisystematic nomenclature and numbering system which received some following, particularly in the German literature. However, many of these names when used contemporaneously with IUPAC names contribute to some of the current confusion in the chemical literature (2). For example, the name 1,2-benzopyrene in the system by Clar refers to what is called in the IUPAC system benzo[e]pyrene, Figure 1; but based on the IUPAC systematic numbering of pyrene, 1,2-benzopyrene refers to the benzo[a]pyrene system, Figure 2.

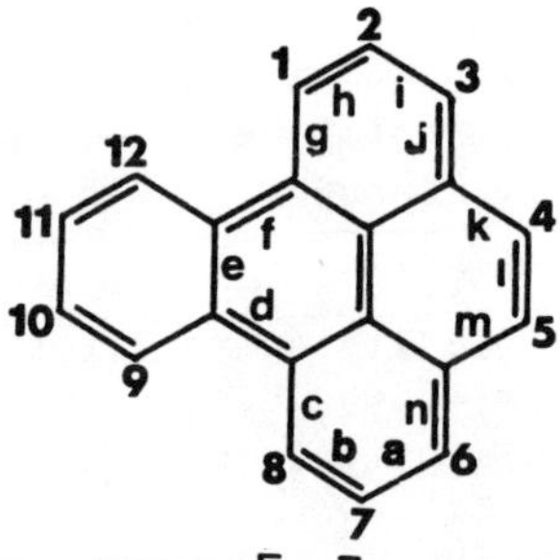

FIGURE 1. BENZO[e]PYRENE

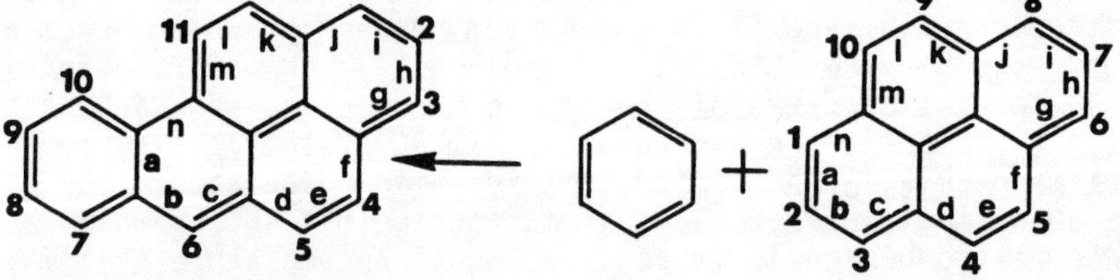

FIGURE 2. BENZO[a]PYRENE

Actually, a great number of synonyms are encountered in the past and present literature of PAHs. Some may even be considered IUPAC synonyms. For example, benzo[def]chrysene may be considered an IUPAC synonym for the preferred IUPAC name benzo[a]pyrene, Figure 2. However, because of the complexity of nomenclature, the use of some of these synonyms has caused glaring errors in the literature. For example, in a recent monograph of the National Academy of Sciences (16) each of two fused systems occur twice in a list but are reported with differing degrees of carcinogenicity. Dibenzo[b,def]-chrysene is reported to be strongly carcinogenic (++) and it is also reported on the same page under a synonymous name, dibenzo[a,h]pyrene, to be even more carcinogenic (+++). Similarly dibenzo[def,p]chrysene is reported to be carcinogenic (+), but under its synonymous name dibenzo[a,l]pyrene its carcinogenicity is reported as uncertain or weakly carcinogenic (±).

Similar inconsistencies can be found in chemical company catalogs. For example, one company (17) lists a compound under its correct IUPAC preferred name, benz[a]anthracene-7,12-dione, but lists on the same page the fundamental ring system from which the compound is formally derived under the old-style name 1,2-benzanthracene, with no cross-reference or synonym for its IUPAC preferred name, benz[a]anthracene. Thus, the exclusive use of the IUPAC preferred fused-ring names and the abandonment of synonyms from older systems is highly recommended.

To aid authors and editors in this effort, a list of approximately 175 chemical synonyms with cross references to the IUPAC preferred names of the 44 polycyclic aromatic hydrocarbons of greatest interest to cancer research is given in Table 1 (Appendix A). These synonyms are taken from the reference file at Chemical Abstracts Service. In addition, to further aid in the understanding of the formation of the IUPAC names, the preferred names, numbering and orientations of the fundamental ring systems are shown in Table 2 (Appendix B).

In Table 2, the ring systems are arranged in order of increasing priorities of the base components (1b). The letter locants of the base components (1c) are shown. For ring systems requiring indicated hydrogen, it is placed arbitrarily at the atom with the lowest possible locant, except for fluorene and carbazole for which the 9H isomers are shown.

The selection of the preferred base component of a ring system generally favors (a) a nitrogenous component if one is present, (b) a heterocyclic component rather than a carbocyclic component, and (c) the component of largest size which has a trivial name. The IUPAC rules (1b) should be consulted for further details. The names of the remaining component rings other than the systematic names cyclopenta, cyclohepta, etc., are converted to ring fusion prefixes by altering the ending of its base name to end in "o". For example, pyrrole becomes pyrrolo, indene becomes indeno. However, acenaphtho, anthra, benzo, naphtho, perylo, phenanthro, furo, imidazo, isoquino, pyrido, pyrimido, quino and thieno are accepted abbreviated prefixes (1d). Also, the final "o" of acenaphtho, benzo, naphtho and perylo and the "a" of the monocyclic prefixes cyclopropa, cyclopenta, cyclohepta, etc., are elided before another vowel, as in benz[a]anthracene. In all other cases, the final "o" or "a" is retained.

The selection of the base component and the naming of component rings is only part of the procedure in naming polycyclic aromatic compounds. In order to name hydrogenated and substituted derivatives, it is necessary to have a single complete numbering system. To achieve this, the fused ring system is oriented by prescribed IUPAC rules (1e). The polycyclic system is oriented so that (a) the greatest number of rings are in a horizontal row and (b) a maximum number of rings are above and to the right of the horizontal row (upper right quadrant). If there is a further choice, the orientation is chosen which has as few rings as possible in the lower left quadrant, Figure 3. Also, it should be emphasized in applying

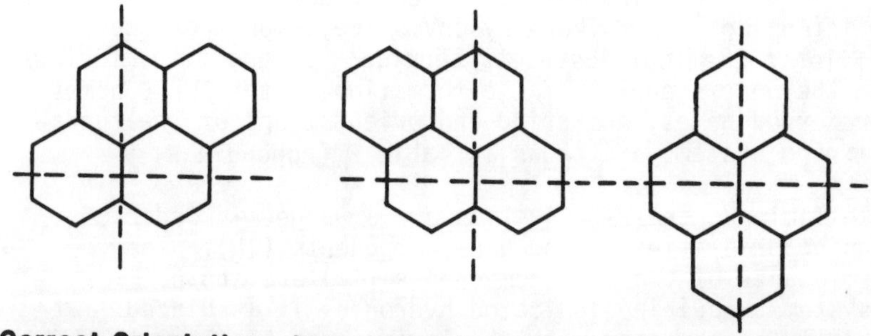

FIGURE 3. ILLUSTRATION OF CORRECT ORIENTATION OF PYRENE

the orientation rules, certain stylized methods for depicting 3, 4, 5, 6, 7, and 8 membered rings shown in Figure 4 have been adopted (1e). In particular, drawing of the six membered ring on its side is avoided if possible, Figure 5.

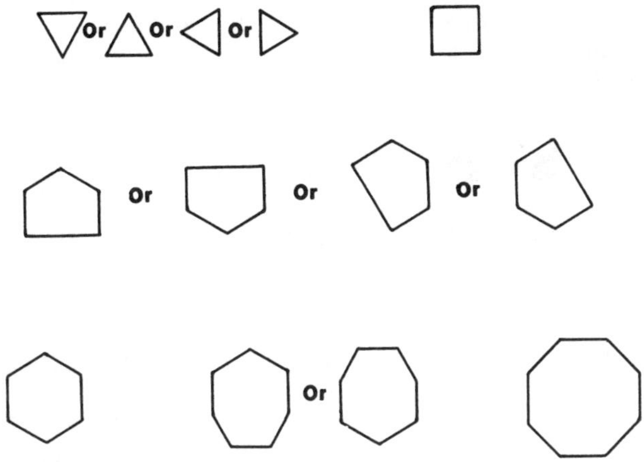

FIGURE 4. STYLIZED DEPICTION OF MONOCYCLES

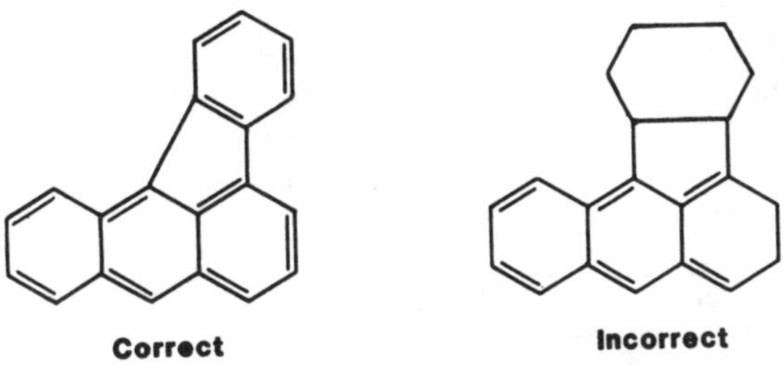

FIGURE 5. ORIENTATION OF BENZ[a]ACEANTHRYLENE

Only after proper orientation of the fused ring system is it given its final numbering. Numbering then begins in a clockwise direction starting with the atom not engaged in ring-fusion in the most counter-clockwise position of the uppermost ring farthest to the right. Carbon atoms common to two or more rings are designated by adding roman letters a, b, c, etc. to the number of the position immediately preceding (1e). In contrast, in compounds containing a heterocyclic base component hetero atoms common to two or more rings are given plain numbers (integers without roman letters). Further fine tuning of the orientation is governed by the criteria of allowing carbon atoms common to two or more rings to follow the lowest possible numbers as illustrated in Figure 6. For heterocycles, further fine tuning of orientation allows lowest possible numbers to be given to hetero atoms. The IUPAC rules should be consulted for details (1f).

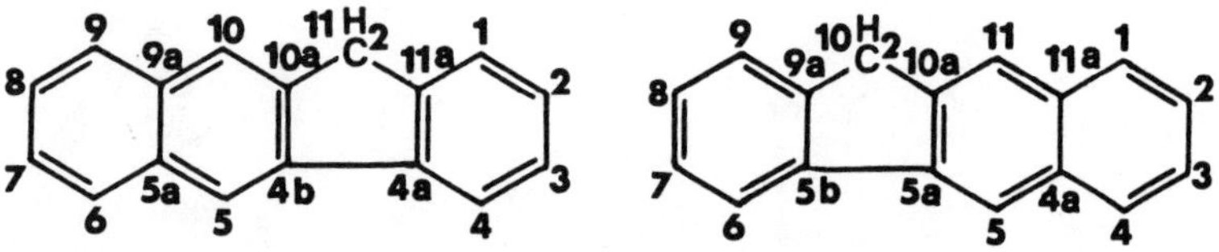

FIGURE 6. THE ORIENTATION AND NUMBERING OF BENZO[b]FLUORENE

Beyond a basic understanding of the rules governing the naming, numbering and orientation of ring systems, perhaps the most useful tool for authors and editors in need of the preferred IUPAC names of fundamental ring systems is the Chemical Abstracts Service Parent Compound Handbook (PCH)(18).[1] It is a comprehensive catalog and index of the thousands of ring systems and natural products which have been reported in the scientific literature and abstracted in Chemical Abstracts (CA) since 1907. Included in the PCH for each ring system is the structural diagram, illustrating the locant numbering system, the CAS Registry Number, the CA index name (usually the IUPAC preferred name), molecular formula, and the CA reference (if the compound was referenced in a CA abstract after Volume 78 [1973]).

Prior to the PCH, its predecessor, The Ring Index (19) with its three supplements served as the catalog of all reported ring systems, and it is still very useful today for its older references and documentation of alternate numberings encountered in the literature through 1963.

It should be emphasized that the preferred IUPAC names for fused ring systems are based on the fundamental form of the ring system containing the maximum number of non-cumulative

[1]The *Ring Systems Handbook* published by the American Chemical Society, Washington, D.C., 1977-present replaces the *Parent Compound Handbook*. It contains structural diagrams for over 72,654 unique ring systems.

double bonds (1a). The systematic fusion prefixes cyclopenta, cyclohepta, cycloocta, etc., denote fusion of the completely unsaturated ring.* Hydrogenated derivatives are named primarily with hydro prefixes (1g) as in Figure 7.

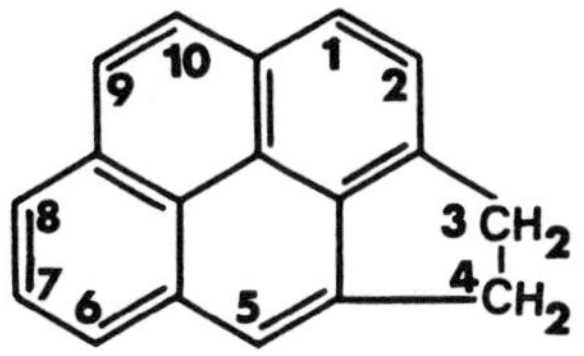

FIGURE 7. 3,4-DIHYDROCYCOPENTA[cd]PYRENE

Methyl substituents are cited as prefixes with appropriate locants as in Figure 8.

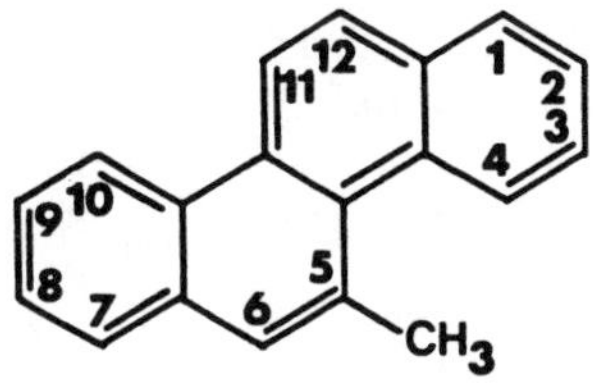

FIGURE 8. 5-METHYLCHRYSENE

Finally, the discussion of the naming, numbering, orientation, and substitution given above is necessarily limited and cannot cover all the complexities involved. Similarly not all

---

* While not sanctioned by IUPAC, fusion prefixes such as cyclopentano have been used for the corresponding hydrogenated derivatives.

of the details of fused-ring nomenclature can be set in black and white in rule form. Thus, for assistance in naming new ring systems and derivatives, authors and editors may consult the Manager of Nomenclature Services, P.O. Box 3012, Columbus, Ohio 43210.

REFERENCES

1. IUPAC (1979): Nomenclature of Organic Chemistry. Sections A, B, C, D, E, F, and H. Pergamon Press, Oxford, 559 pp. (a) rule A-21.3, (b) rules A-21, B-2 and B-3, (c) rule A-21.5, (d) rules A-21.4 and B-3.3, (e) rule A-22, (f) rule B-3.4, (g) rule A-23.
2. Jacob, J. and Grimmer, G. (1977): Nomenclature and structure of polycyclic aromatic hydrocarbons. Int. Agency Res. Cancer, 3(29):15-30.
3. Acheson, R.M. (1973): Nomenclature and numbering system. Chem. Heterocycl. Compounds, 9:1-8.
4. McNaught, A.D. (1976): The nomenclature of heterocycles. Advan. Heterocycl. Chem., 20:175-247.
5. Hata, K. (1979): The IUPAC nomenclature of polycyclic compounds. Yuki Gosei Kagaku Kyokaishi, 37(2):148-160.
6. Balaban, A.T. (1976): Enumeration of cyclic graphs. In: Chemical Applications of Graph Theory. Academic Press, London, Chapter 5.
7. Patterson, A.M. (1925): Proposed international rules for numbering organic ring systems. J. Am. Chem. Soc., 47:543-561.
8. Patterson, A.M. (1928): The nomenclature of parent ring systems. J. Am. Chem. Soc., 50:3074-3087.
9. Chemical Abstracts Service (1982): Naming and indexing of chemical substances for Chemical Abstracts. CA Index Guide, Appendix IV, Sections 147-153.
10. Balaban, A.T., and Harary, F. (1968): Enumeration and proposed nomenclature of benzenoid cata-condensed polycyclic aromatic hydrocarbons. Tetrahedron, 24:2505-2516.
11. Bonchev, D., and Balaban, A.T. (1981): Topological centric coding and nomenclature of polycyclic hydrocarbons. 1. Condensed benzenoid systems (polyhexes, fusenes). J. Chem. Inf. Comput. Sci., 21:223-229.
12. Lozac'h, N. Goodson, A.L., and Powell, W.H. (1979): Nodal nomenclature, general principles, Angew. Chem. Intern. Ed., 18(12):887-899.

13. Bally, O., and Scholl, R. (1911): Action of glycerol and H_2SO_4 on amino compounds and on substances free from N in the anthracene series; benzanthrone and its reduction products, nomenclature. Ber., 44:1662-1665.
14. Stelzner, R. and Kuh, H. (1921): Nomenklatur-Fragen, Lit. Reg. Org. Chem., 3:1-100.
15. Clar, E. (1964): Nomenclature of polycyclic hydrocarbons. In: Polycyclic Hydrocarbons, Volume 1, Chapter 1, pp. 1-11, by E. Clar, Academic Press, London.
16. National Research Council. Committee on Biologic Effects of Atmospheric Pollutants (1972): Structure and nomenclature of polycyclic aromatic hydrocarbons and aza-arenes. In: Particulate Polycyclic Organic Matter, Biological Effects of Atmospheric Pollutants. Chapter 2, p. 4-12. National Academy of Sciences, Washington, D.C.
17. Aldrich Catalog/Handbook of Fine Chemicals (1982-1983). Milwaukee, Wisconsin.
18. The Parent Compound Handbook (1976-   ): Chemical Abstracts Service, Columbus, Ohio.
19. The Ring Index (1960): 2nd Edition, edited by A. M. Patterson, L. T. Capell, and D. F. Walker, American Chemical Society, Washington, D.C. Supplement I (1963), Supplement II (1964), Supplement III (1965).

APPENDIX A - TABLE 1. SYNONYM AND IUPAC NAME LIST

ALPHABETIZED PAH SYNONYMS	IUPAC PREFERRED NAMES
Acepyrene	see Cyclopenta[cd]pyrene
Acepyrylene	see Cyclopenta[cd]pyrene
Anthanthren	see Dibenzo[def,mno]chrysene
Anthanthrene	see Dibenzo[def,mno]chrysene
Anthracene	IUPAC preferred name. Table 2, structure 7.
Anthracin	see Anthracene
7-Azabenz[a]anthracene	see Benz[a]acridine
12-Azabenz[a]anthracene	see Benz[c]acridine
7-Azadibenz[a,h]anthracene	see Dibenz[a,h]acridine
7-Azadibenz[a,j]anthracene	see Dibenz[a,j]acridine
7-Aza-7H-dibenzo[c,g]fluorene	see 7H-Dibenzo[c,g]carbazole
9-Azafluorene	see Carbazole
Ba 51-090462	see Dibenzo[def,p]chrysene
Benz[e]acephenanthrylene	IUPAC preferred name. Table 2, structure 16.
3,4-Benz[e]acephenanthrylene	see Benz[e]acephenanthrylene
Benz[a]acridine	IUPAC preferred name. Table 2, structure 39.
Benz[c]acridine	IUPAC preferred name. Table 2, structure 40.
1,2-Benzacridine	see Benz[a]acridine
3,4-Benzacridine	see Benz[c]acridine
Benzanthracene	see Benz[a]anthracene
Benz[a]anthracene	IUPAC preferred name. Table 2, structure 8.
1,2-Benzanthracene	see Benz[a]anthracene
1,2-Benz[a]anthracene	see Benz[a]anthracene
1,2:5,6-Benzanthracene	see Dibenz[a,h]anthracene
Benzanthrene	see Benz[a]anthracene
1,2-Benzanthrene	see Benz[a]anthracene
Benzfluoranthene	see Benz[e]acephenanthrylene
2,3-Benzfluoranthene	see Benz[e]acephenanthrylene
3,4-Benzfluoranthene	see Benz[e]acephenanthrylene
7,8-Benzfluoranthene	see Benzo[j]fluoranthene
8,9-Benzfluoranthene	see Benzo[k]fluoranthene
3,4-Benzoacridine	see Benz[c]acridine
Benzoanthracene	see Benz[a]anthracene
Benzo[a]anthracene	see Benz[a]anthracene
1,2-Benzoanthracene	see Benz[a]anthracene

Benzo[def]chrysene	see Benzo[a]pyrene
Benzo-12,13-fluoranthene	see Benzo[j]fluoranthene
Benzo[b]fluoranthene	see Benz[e]acephenanthrylene
Benzo[e]fluoranthene	see Benz[e]acephenanthrylene
Benzo[ghi]fluoranthene	IUPAC preferred name. Table 2, structure 14.
Benzo[j]fluoranthene	IUPAC preferred name. Table 2, structure 12.
Benzo[k]fluoranthene	IUPAC preferred name. Table 2, structure 13.
Benzo[l]fluoranthene	see Benzo[j]fluoranthene
Benzo[mno]fluoranthene	see Benzo[ghi]fluoranthene
2,3-Benzofluoranthene	see Benz[e]acephenanthrylene
2,13-Benzofluoranthene	see Benzo[ghi]fluoranthene
3,4-Benzofluoranthene	see Benz[e]acephenanthrylene
7,10-Benzofluoranthene	see Benzo[ghi]fluoranthene
8,9-Benzofluoranthene	see Benzo[k]fluoranthene
10,11-Benzofluoranthene	see Benzo[j]fluoranthene
11,12-Benzofluoranthene	see Benzo[k]fluoranthene
2,3-Benzofluoranthrene	see Benzo[e]acephenanthrylene
1H-Benzo[a]fluorene	IUPAC preferred name. Table 2, structure 2.
1H-Benzo[b]fluorene	IUPAC preferred name. Table 2, structure 3.
1H-Benzo[c]fluorene	IUPAC preferred name. Table 2, structure 4.
Benzo[jk]fluorene	see Fluoranthene
2,3-Benzofluorene	see Benzo[b]fluorene
Benzo[rst]pentaphene	IUPAC preferred name. Table 2, structure 34.
Benzo[ghi]perylene	IUPAC preferred name. Table 2, structure 32.
1,12-Benzoperylene	see Benzo[ghi]perylene
Chrysene	IUPAC preferred name. Table 2, structure 26.
Benzo[a]phenanthrene	see Chrysene
Benzo[b]phenanthrene	see Benz[a]anthracene
Benzo[c]phenanthrene	IUPAC preferred name. Table 2, structure 6.
Benzo[def]phenanthrene	see Pyrene
Benzo[l]phenanthrene	see Triphenylene
1,2-Benzophenanthrene	see Chrysene
2,3-Benzophenanthrene	see Benz[a]anthracene
3,4-Benzophenanthrene	see Benzo[c]phenanthrene
9,10-Benzophenanthrene	see Triphenylene

Benzopyrene — see Benzo[*a*]pyrene

Benzo[*a*]pyrene — IUPAC preferred name. Table 2, structure 22.

Benzo[*e*]pyrene — IUPAC preferred name. Table 2, structure 23.

1,2-Benzopyrene — see Benzo[*a*]pyrene or Benzo[*e*]pyrene

3,4-Benzopyrene — see Benzo[*a*]pyrene

6,7-Benzopyrene — see Benzo[*a*]pyrene

Benzo[*b*]triphenylene — IUPAC preferred name. Table 2, structure 20.

1,12-Benzperylene — see Benzo[*ghi*]perylene

1,2-Benzphenanthrene — see Chrysene

9,10-Benzphenanthrene — see Triphenylene

Benz[*a*]pyrene — see Benzo[*a*]pyrene

1,2-Benzpyrene — see Benzo[*a*]pyrene or Benzo[*e*]pyrene

3,4-Benzpyrene — see Benzo[*a*]pyrene

4,5-Benzpyrene — see Benzo[*a*]pyrene

3,4-Benz[*a*]pyrene — see Benzo[*a*]pyrene

2,3:1',8'-Binaphthylene — see Benzo[*k*]fluoranthene

o-Biphenylenemethane — see Fluorene

BP — see Benzo[*a*]pyrene

Carbazole — IUPAC preferred name. Table 2. structure 36.

9*H*-Carbazole — see also Carbazole

Coronene — IUPAC preferred name. Table 2, structure 35.

Cyclopentano[*cd*]pyrene — see 3,4-Dihydrocyclopenta[*cd*]pyrene

Cyclopenta[*cd*]pyrene — IUPAC preferred name. Table 2, structure 24.

Cyclopenteno[*cd*]pyrene — see Cyclopenta[*cd*]pyrene

DBA — see Dibenz[*a,h*]anthracene

Dibenz[*a,e*]aceanthrylene — IUPAC preferred name. Table 2, structure 18.

Dibenz[*a,d*]acridine — see Dibenz[*a,h*]acridine

Dibenz[*a,f*]acridine — see Dibenz[*a,j*]acridine

Dibenz[*a,h*]acridine — IUPAC preferred name. Table 2, structure 41.

Dibenz[*ah*]acridine — see Dibenz[*a,h*]acridine

Dibenz[*a,j*]acridine — IUPAC preferred name. Table 2, structure 42.

1,2,5,6-Dibenzacridine — see Dibenz[*a,h*]acridine

1,2:5,6-Dibenzacridine — see Dibenz[*a,h*]acridine

1,2,7,8-Dibenzacridine — see Dibenz[*a,j*]acridine

1,2:7,8-Dibenzacridine — see Dibenz[*a,j*]acridine

Dibenz[*de,kl*]anthracene — see Perylene

1,2:3,4-Dibenzanthracene	see Benzo[b]triphenylene
1,2:5,6-Dibenzanthracene	see Dibenz[a,h]anthracene
1,2:5,6-Dibenz[a]anthracene	see Dibenz[a,h]anthracene
1,2,7,8-Dibenzanthracene	see Dibenz[a,j]anthracene
Dibenz[a,c]anthracene	see Benzo[b]triphenylene
Dibenz[a,h]anthracene	IUPAC preferred name. Table 2, structure 9.
Dibenz[a,j]anthracene	IUPAC preferred name. Table 2, structure 10
3,4,5,6-Dibenzanthracene	see Dibenz[a,j]anthracene
1,2,3,4-Dibenznaphthalene	see Triphenylene
Dibenzo[a,j]acridine	see Dibenz[a,j]acridine
1,2,5,6-Dibenzoacridine	see Dibenz[a,h]acridine
Dibenzo[a,c]anthracene	see Benzo[b]triphenylene
Dibenzo[a,h]anthracene	see Dibenz[a,h]anthracene
1,2:3,4-Dibenzoanthracene	see Benzo[b]triphenylene
1,2,5,6-Dibenzoanthracene	see Dibenz[a,h]anthracene
Dibenzo-1,2,7,8-anthracene	see Dibenz[a,j]anthracene
7H-Dibenzo[c,g]carbazole	IUPAC preferred name. Table 2, structure 37.
3,4,5,6-Dibenzocarbazole	see 7H-Dibenzo[c,g]carbazole
Dibenzo[b,def]chrysene	IUPAC preferred name. Table 2, structure 27.
Dibenzo[def,mno]chrysene	IUPAC preferred name. Table 2, structure 28.
Dibenzo[def,p]chrysene	IUPAC preferred name. Table 2, structure 29.
Dibenzo[a,e]fluoranthene	see Dibenz[a,e]aceanthrylene
2,3,5,6-Dibenzofluoranthene	see Dibenz[a,e]aceanthrylene
Dibenzo[a,jk]fluorene	see Benzo[j]fluoranthene
Dibenzo[b,jk]fluorene	see Benzo[k]fluoranthene
Dibenzo[a,e]pyrene	see Naphtho[1,2,3,4-def]chrysene
Dibenzo[a,h]pyrene	see Dibenzo[b,def]chrysene
Dibenzo[a,i]pyrene	see Benzo[rst]pentaphene
Dibenzo[a,l]pyrene	see Dibenzo[def,p]chrysene
Dibenzo[b,h]pyrene	see Benzo[rst]pentaphene
Dibenzo[cd,jk]pyrene	see Dibenzo[def,mno]chrysene
1,2:3,4-Dibenzopyrene	see Dibenzo[def,p]chrysene
1,2:4,5-Dibenzopyrene	see Naphtho[1,2,3,4-def]chrysene
1,2,9,10-Dibenzopyrene	see Dibenzo[def,p]chrysene
3,4:8,9-Dibenzopyrene	see Dibenzo[b,def]chrysene
3,4:9,10-Dibenzopyrene	see Benzo[rst]pentaphene
Dibenzopyrrole	see Carbazole
Dibenzo[b,d]pyrrole	see Carbazole
1,2:7,8-Dibenzpyrene	see Benzo[rst]pentaphene
2,3:4,5-Dibenzpyrene	see Dibenzo[def,p]chrysene
4,5,6,7-Dibenzpyrene	see Dibenzo[def,p]chrysene

3,4-Dihydrocyclopenta[*cd*]pyrene	IUPAC preferred name. See Figure
3,4-Dimethylenepyrene	see 3,4-Dihydrocyclopenta[*cd*]pyren
1,4-Dimethylphenanthrene	IUPAC preferred name.
peri-Dinaphthalene	see Perylene
Diphenylememethane	see Fluorene
Diphenylenimine	see Carbazole
Fluoranthene	IUPAC preferred name. Table 2. structure 11.
Fluorene	IUPAC preferred name. Table 2, structure 1.
9*H*-Fluorene	see Fluorene
Green Oil	see Anthracene
1,2,5,6,9,10-Hexahydrocoronene	see Coronene
Idryl	see Fluoranthene
Indeno[1,2,3-*cd*]pyrene	IUPAC preferred name. Table 2. structure 25.
Isochrysene	see Triphenylene
2-Methylchrysene	IUPAC preferred name.
5-Methylchrysene	IUPAC preferred name. See Figure
2,2'-Methylenebiphenyl	see Fluorene
2-Methylfluoranthene	IUPAC preferred name.
3-Methylfluoranthene	IUPAC preferred name.
4-Methylfluoranthene	IUPAC preferred name.
1-Methylphenanthrene	IUPAC preferred name.
1,2-(1,8-Naphthalenediyl)benzene	see Fluoranthene
Naphtho[1,2,3,4-*def*]chrysene	IUPAC preferred name. Table 2, structure 30.
1,2-(1,8-Naphthylene)benzene	see Fluoranthene
Paranaphthalene	see Anthracene
Perilene	see Perylene
Perylene	IUPAC preferred name. Table 2, structure 31.
Phenanthrene	IUPAC preferred name. Table 2, structure 6.
o-Phenylenepyrene	see Indeno[1,2,3-*cd*]pyrene
1,10-(*o*-Phenylene)pyrene	see Indeno[1,2,3-*cd*]pyrene
1,10-(1,2-Phenylene)pyrene	see Indeno[1,2,3-*cd*]pyrene
Pyrene	IUPAC preferred name. Table 2, structure 21.
β-Pyrene	see Pyrene
Tetrahelicene	see Benzo[*c*]phenanthrene
Tetra Olive N2G	see Anthracene
Tetraphene	see Benz[*a*]anthracene
Triphenylene	IUPAC preferred name. Table 2, structure 19.

APPENDIX B - TABLE 2. PAH STRUCTURES

PREFERRED IUPAC NAMES, ORIENTATION AND NUMBERING OF SOME POLYCYCLIC AROMATIC HYDROCARBONS AND HETEROCYCLES

1.

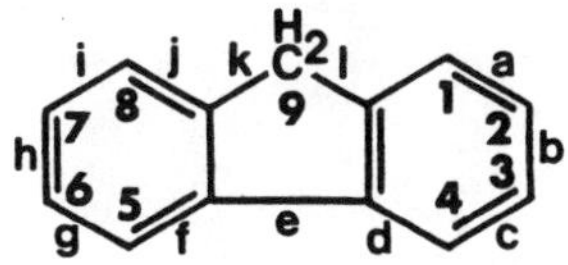

FLUORENE OR 9<u>H</u>-FLUORENE

2.

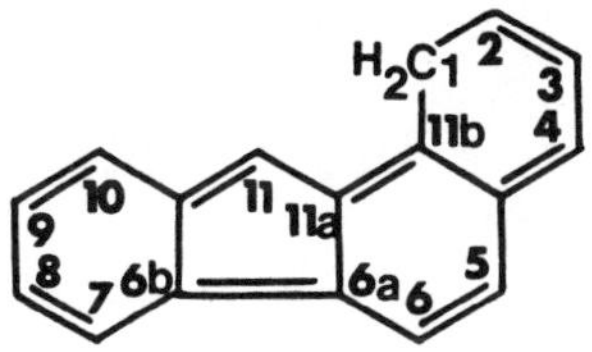

1<u>H</u>-BENZO[<u>a</u>]FLUORENE

3.

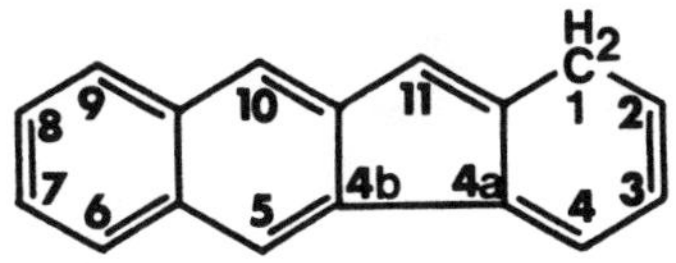

1<u>H</u>-BENZO[<u>b</u>]FLUORENE

4.

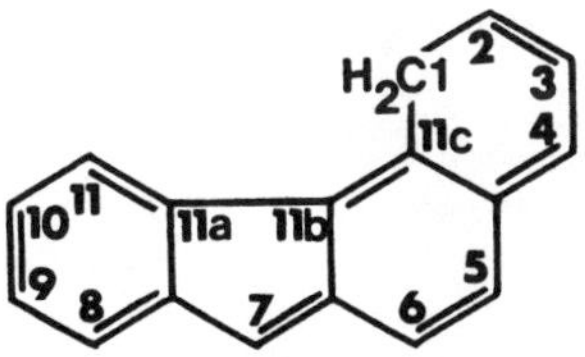

1<u>H</u>-BENZO[<u>c</u>]FLUORENE

5.

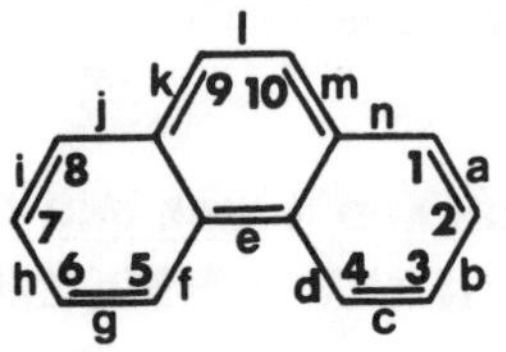

✱ PHENANTHRENE

6.

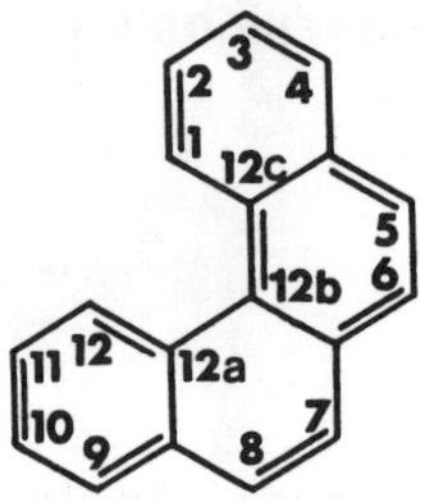

BENZO[c]PHENANTHRENE

7.

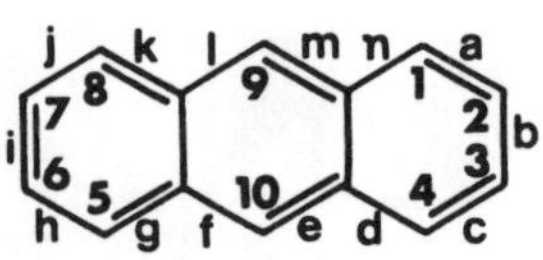

✱ ANTHRACENE

8.

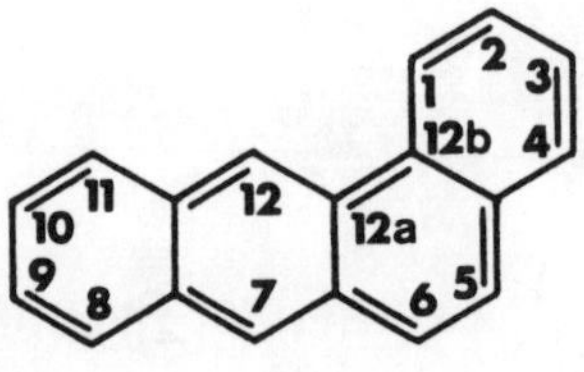

BENZ[a]ANTHRACENE

9.

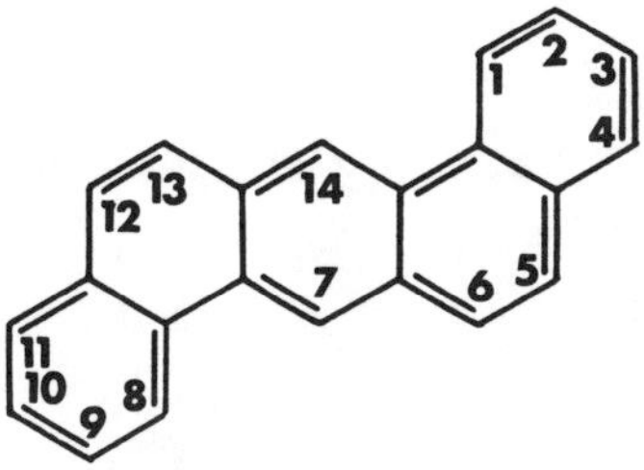

DIBENZ[a,h]ANTHRACENE

10.

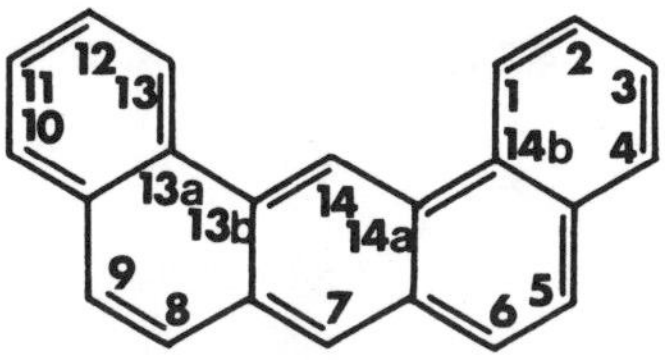

DIBENZ[a,j]ANTHRACENE

11.

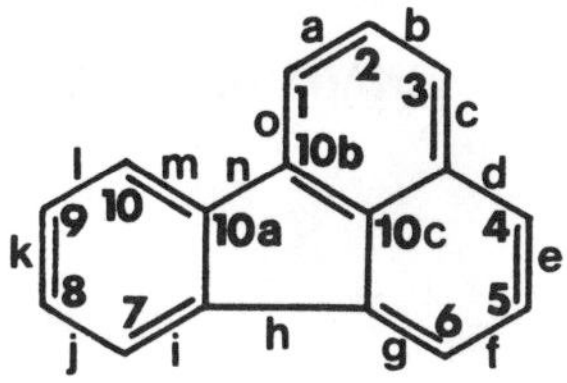

FLUORANTHENE

12.

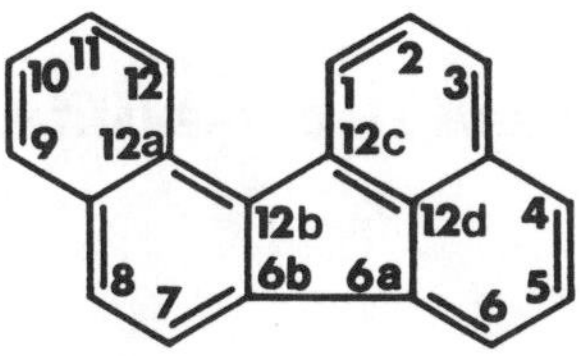

BENZO[j]FLUORANTHENE

13.

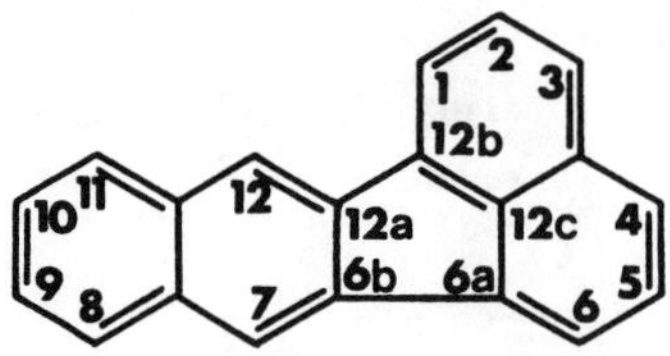

BENZO[k]FLUORANTHENE

14.

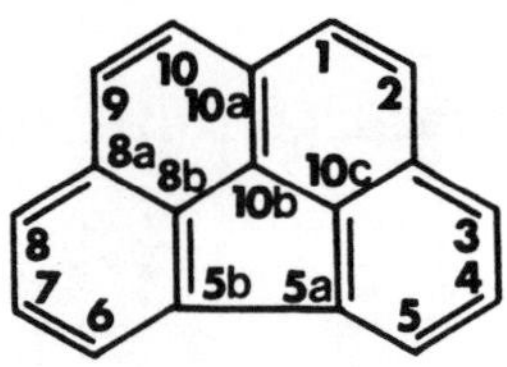

BENZO[ghi]FLUORANTHENE

15.

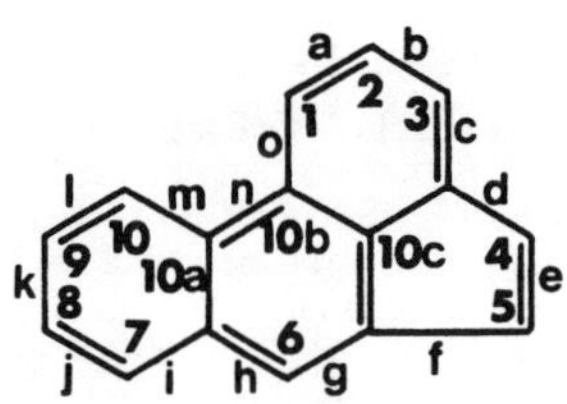

ACEPHENANTHRYLENE

16.

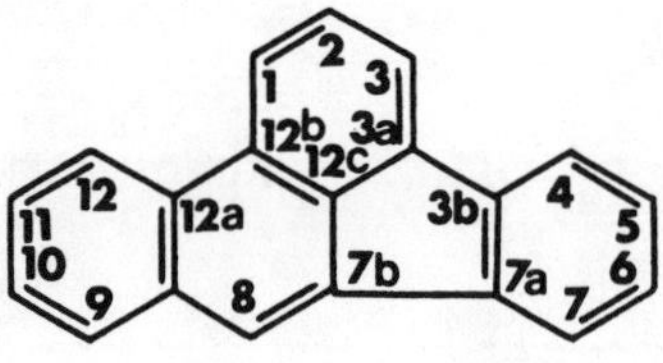

BENZ[e]ACEPHENANTHRYLENE

17.

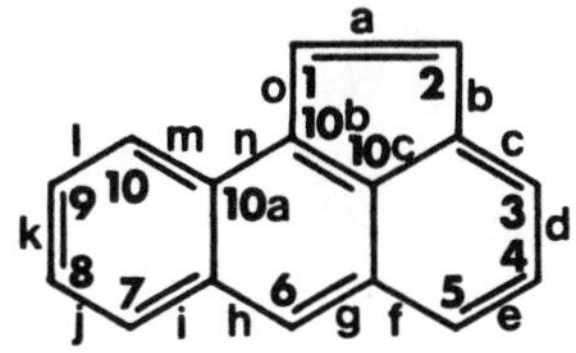

ACEANTHRYLENE

18.

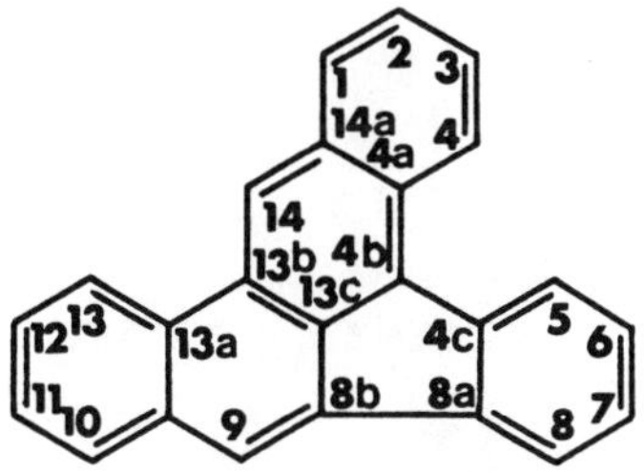

DIBENZ[a,e]ACEANTHRYLENE

19.

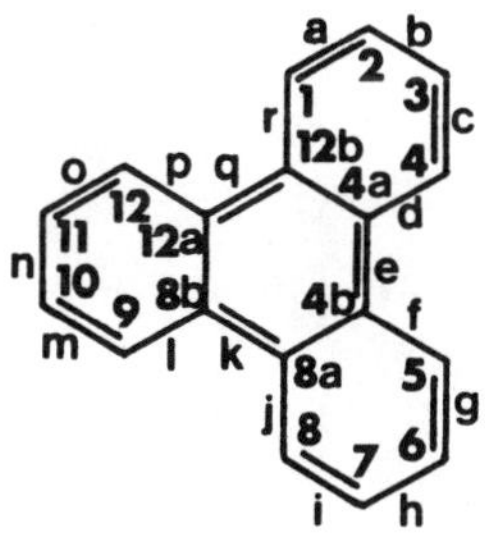

TRIPHENYLENE

20.

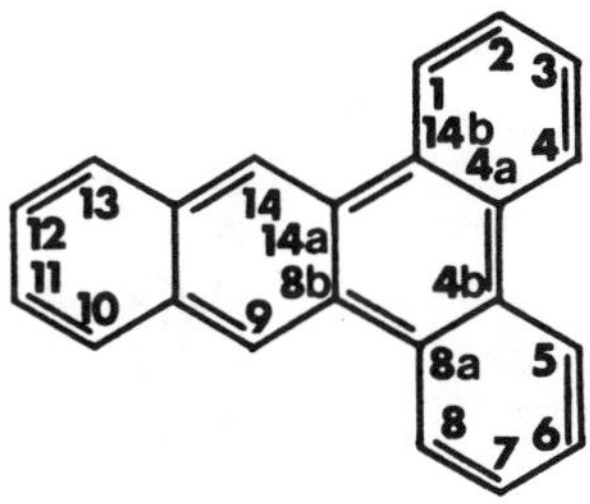

BENZO[b]TRIPHENYLENE

21.

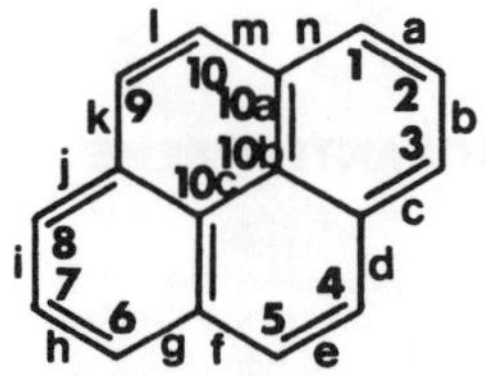

PYRENE

22.

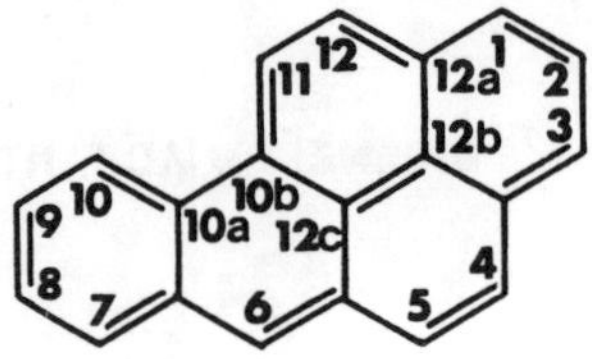

BENZO[a]PYRENE

23.

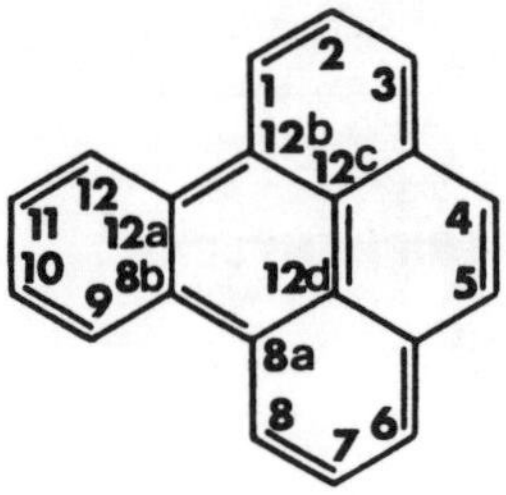

BENZO[e]PYRENE

24.

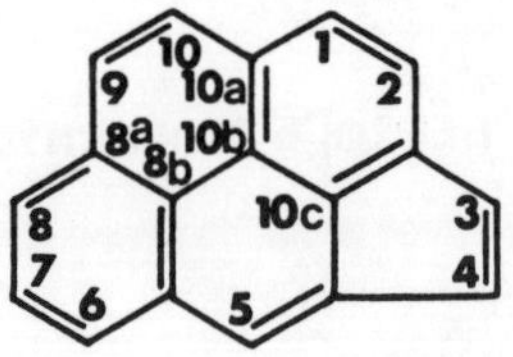

CYCLOPENTA[cd]PYRENE

25.

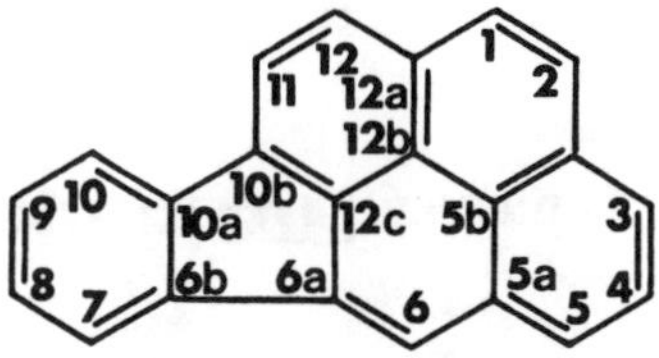

INDENO[1,2,3-cd]PYRENE

26.

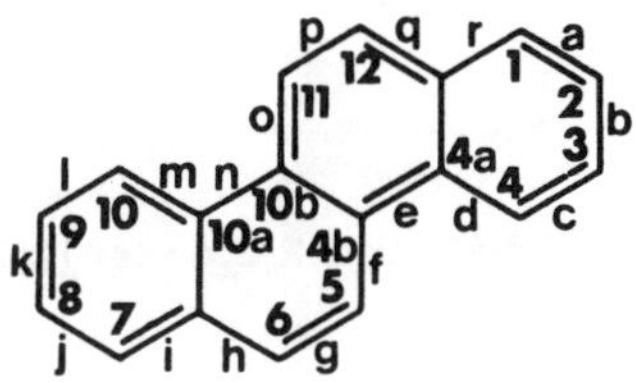

CHRYSENE

27.

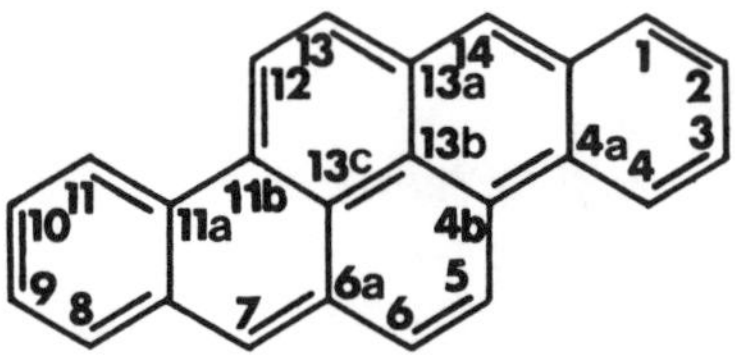

DIBENZO[b,def]CHRYSENE

28.

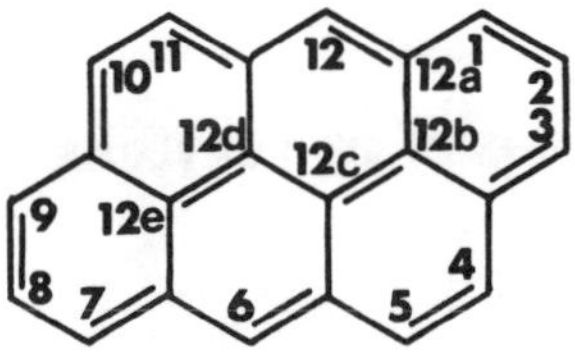

DIBENZO[def,mno]CHRYSENE

29.

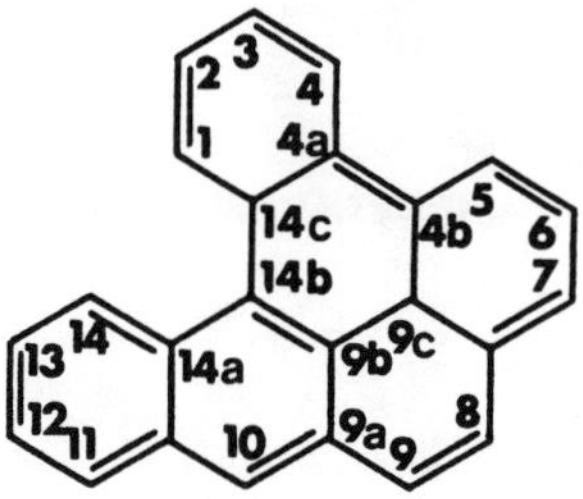

DIBENZO[def,p]CHRYSENE

30.

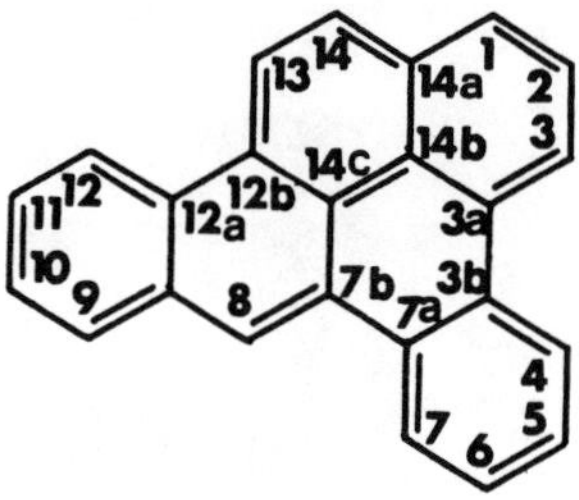

NAPHTHO[1,2,3,4-def]CHRYSENE

31.

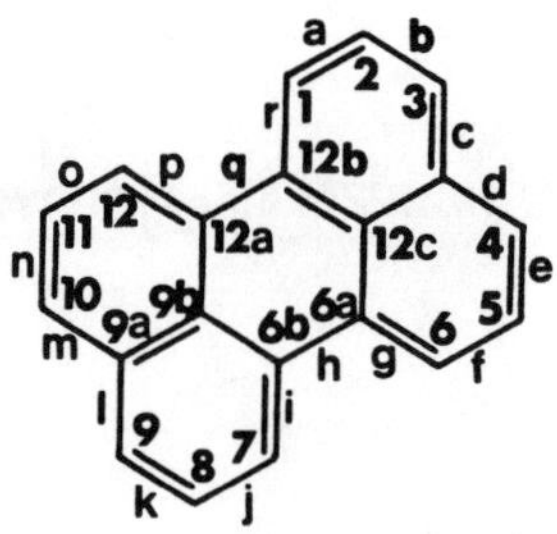

PERYLENE

32.

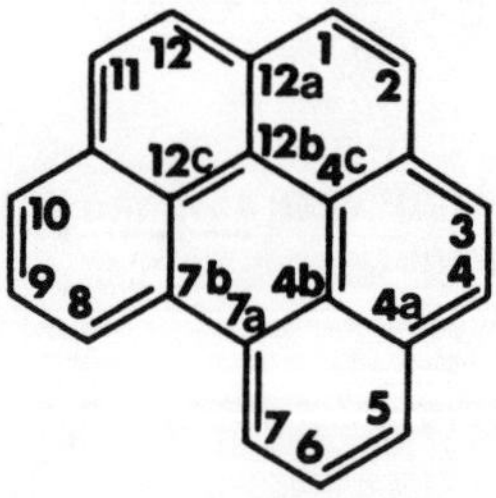

BENZO[ghi]PERYLENE

33.

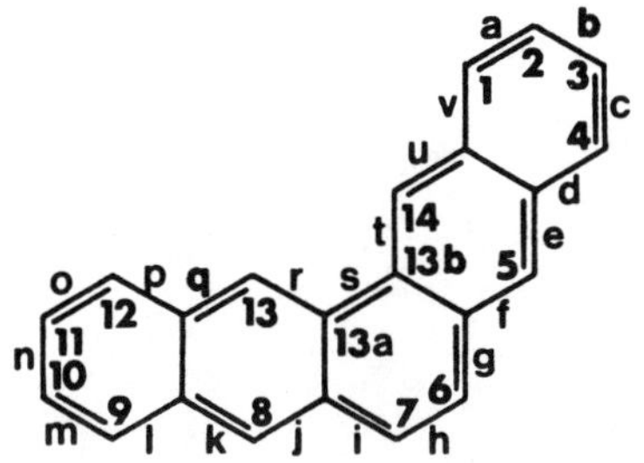

PENTAPHENE

34.

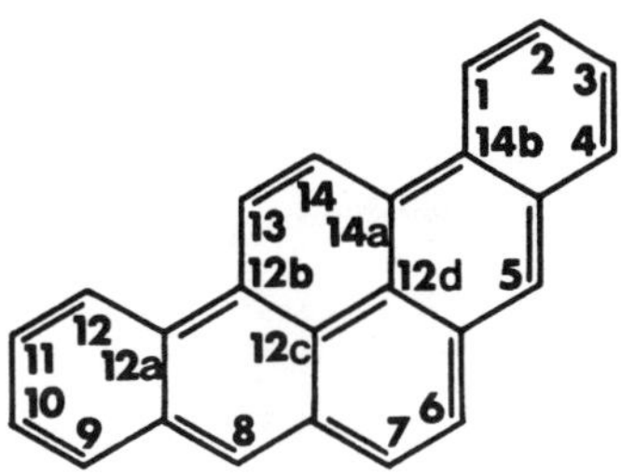

BENZO[rst]PENTAPHENE

35.

CORONENE

36.

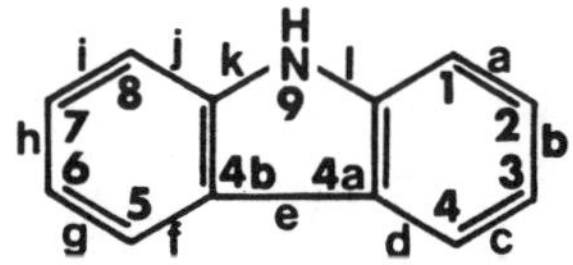

CARBAZOLE
OR
*** 9H-CARBAZOLE**

37.

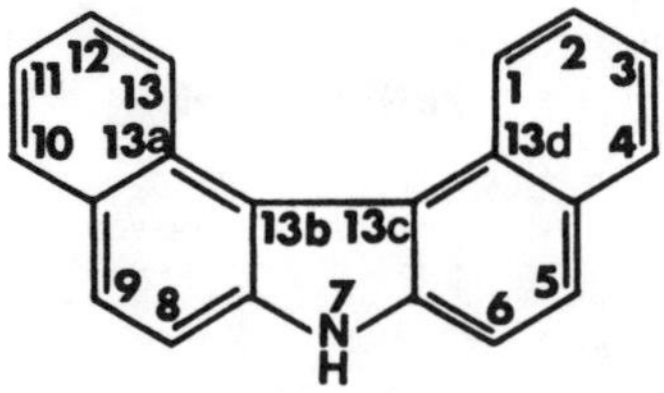

7H-DIBENZO[c,g]CARBAZOLE

38.

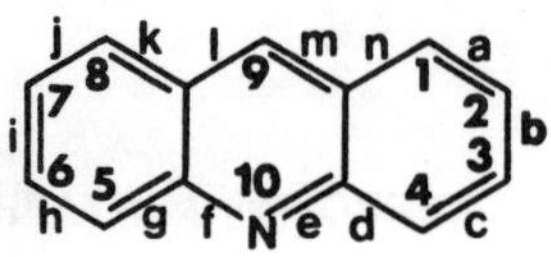

* ACRIDINE

39.

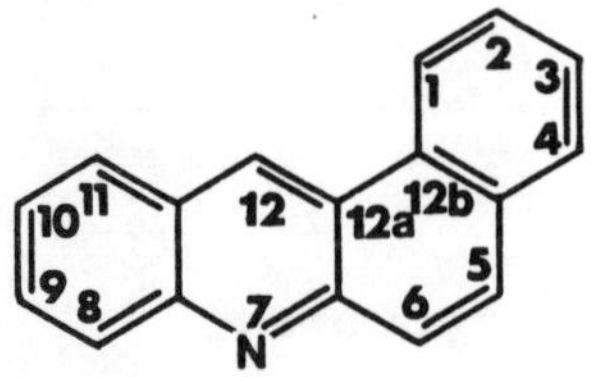

BENZ[a]ACRIDINE

40.

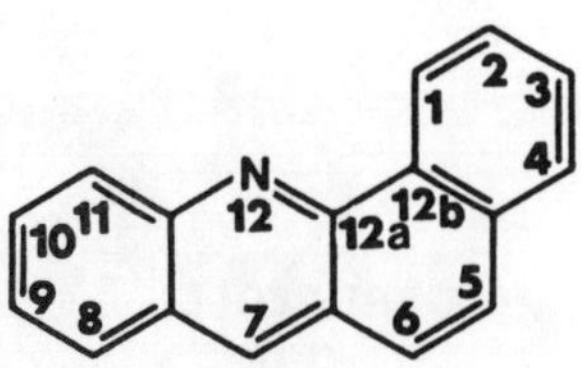

BENZ[c]ACRIDINE

41. 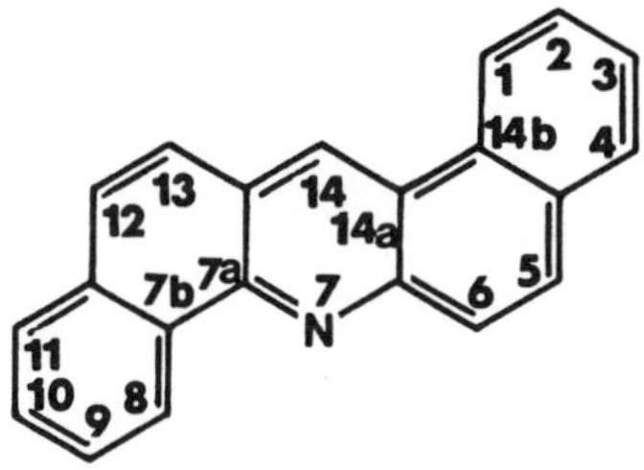 BENZ[a,h]ACRIDINE

42. 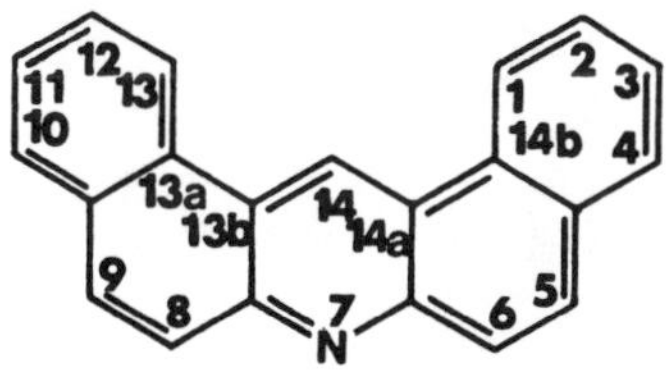 DIBENZ[a,j]ACRIDINE

---

* Denotes exception to systematic numbering

Chapter 2

SYSTEMATIC NOMENCLATURE OF THE NITROGEN, OXYGEN, AND SULFUR FUNCTIONAL POLYCYCLIC AROMATIC COMPOUNDS

DOUGLAS W. LATER, CHERYLYN W. WRIGHT,
KURT L. LOENING AND JOY E. MERRITT

INTRODUCTION

Heterocyclic systems include fused-ring compounds containing atoms of at least two different elements as ring members. Nitrogen and carbon are the two elements for the nitrogen-containing polycyclic aromatic compounds (N-PAC); oxygen and carbon are the two elements for the oxygen-containing polycyclic aromatic compounds (O-PAC); likewise, sulfur and carbon are the two elements for the sulfur-containing polycyclic aromatic compounds (S-PAC). In addition, nitrogen, oxygen and sulfur functional groups exocyclic to fused-ring systems will be considered in this paper as members of the N-, O- and S-PAC groups, respectively. (For the purposes of this paper, fused-ring compounds with the heteroatom as a ring member are defined as endocyclic and heteroatoms positioned as a functional group exterior to the ring are defined as exocyclic.)

The nomenclature used in the literature for the heterocyclic and exocyclic PAC is often diversified and confusing. Different preferences for naming generic classes and individual heterofunctional PAC are common. An increase in the number of papers presented in the PAH symposium proceedings on N-, O- and S-PAC has prompted a brief compilation of the key rules of nomenclature for these classes of compounds. Hence, a synopsis of the current rules for nomenclature applied to monofunctional N-, O-, and S-PAC as published by the International Union of Pure and Applied Chemistry (IUPAC) [1] is presented here in an effort to clarify and unify nomenclature used in the proceedings of these symposia. The nomenclature methods that are abridged and presented in this paper were taken mainly from Sections A, B, and C of the IUPAC rules on nomenclature of organic compounds. Other summaries of these rules, as they apply to polycyclic aromatic hydrocarbon (PAH) chemistry, have also been published and are recommended supplementary information [2-7]. Applications of proper nomenclature for known fused-ring compounds have also been compiled [7,8].

In this paper, the nomenclature for the nonfunctional PAH will first be briefly reviewed, followed by three sections devoted to outlining the key aspects of naming the N-, O- and S-PAC, respectively. Encouragement is offered to investigators and authors in this field of research to use the IUPAC rules of nomenclature in their work and publications.

PAH NOMENCLATURE REVIEW

Fundamentals of nomenclature for the nonfunctional PAH that are applicable to naming the nitrogen-, oxygen- and sulfur-containing PAC are found mostly in Sections A21 and A22 of the IUPAC rules. Rules of PAH nomenclature will not be reviewed in detail in this section. However, since the PAH nomenclature is fundamental to N-, O-, and S-PAC nomenclature, a few key principles will be mentioned. A listing of retained trivial names of selected PAH that are particularly applicable are shown in Table 1. Also shown are the accepted numbering schemes and radical names for each of the PAH. Fused-ring PAH that have no accepted trivial name are named by prefixing the names of the simplest component ring systems and joining them with the name of the base component which contains as many rings as possible and occurs as high in Table 1 as possible (IUPAC Rule A21.3 and A21.4). The prefixes designating attached PAH components to the base component are formed by changing the ending "-ene" of the name of the component PAH to "-eno", with the exceptions listed in Table 2. When more than one prefix is present, they are arranged in alphabetical order (IUPAC Rule A21.4).

Spatial orientation of the polycyclic system requires (a) the greatest number of rings in a horizontal row, (b) a maximum number of rings above and to the right of the horizontal row (upper right-hand quadrant), and (c) as few rings as possible in the lower left-hand quadrant, (IUPAC Rule A22.1). Numbering, in most cases, begins with the first carbon atom not engaged in ring fusion in the upper right-hand ring of the top row and proceeds clockwise around the molecule (IUPAC Rule A22.1). There are, however, notable exceptions such as anthracene, phenanthrene and others (see Table 1). Fused-ring isomers are denoted by lettering the

TABLE 1

IUPAC STRUCTURES, NUMBERING, AND TRIVIAL NAMES FOR SELECTED PAH

IUPAC No.[a]	Structure[b]	Trivial Name	Radical Name[b]
(2)	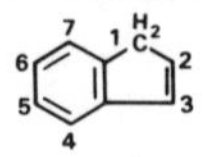	indene or 1*H*-indene[c]	indenyl- or 1*H*-indenyl-
(3)	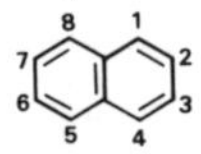	naphthalene	naphthyl-*
(9)	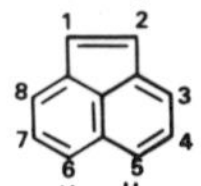	acenaphthylene	acenaphthyl-*
---	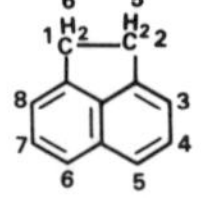	acenaphthene	acenaphthenyl-
---	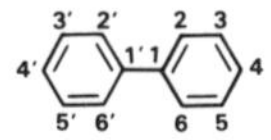	biphenyl	biphenylyl-
(10)	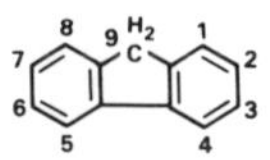	fluorene or 9*H*-fluorene[c]	fluorenyl- or 9*H*-fluorenyl-
(12)	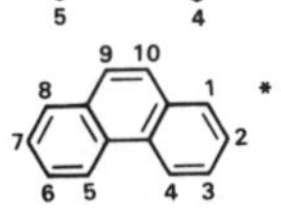	phenanthrene	phenanthryl-*
(13)	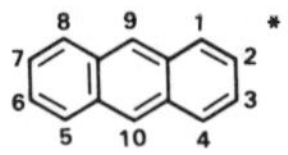	anthracene	anthryl-*
(14)	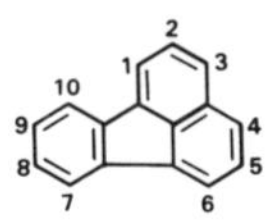	fluoranthene	fluoranthenyl-
(15)	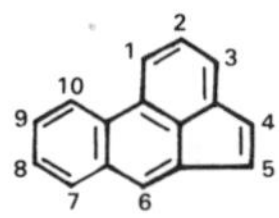	acephenanthrylene	acephenanthryl-*
(16)	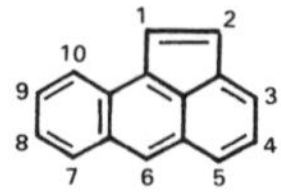	aceanthrylene	aceanthryl-*

[a] Listed in reverse order of preference
[b] *Indicates exception to numbering and/or radical naming rules
[c] Currently used in Chemical Abstracts

TABLE 1 (CONTINUED)

IUPAC No.[a]	Structure[b]	Trivial Name	Radical Name[b]
(17)	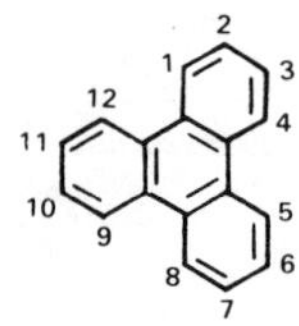	triphenylene	triphenylenyl-
(18)	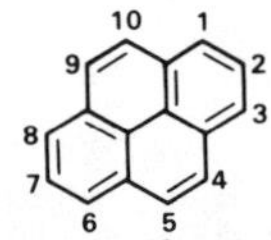	pyrene	pyrenyl-
(19)	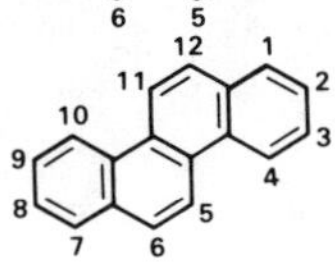	chrysene	chrysenyl-
(20)	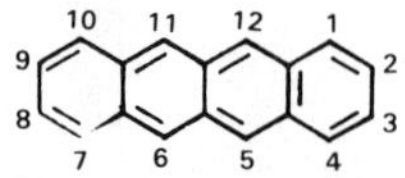	naphthacene	naphthacenyl-
(22)	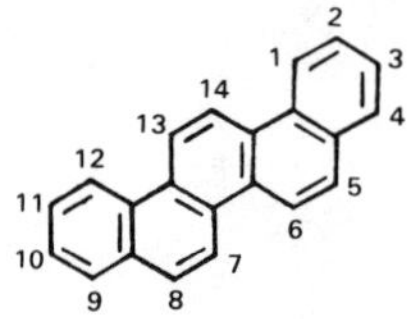	picene	picenyl-
(23)	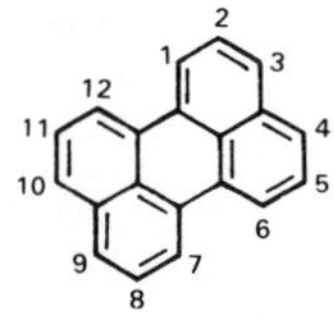	perylene	perylenyl-
(24)	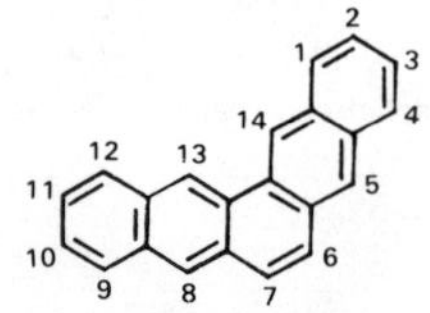	pentaphene	pentaphenyl-
(25)	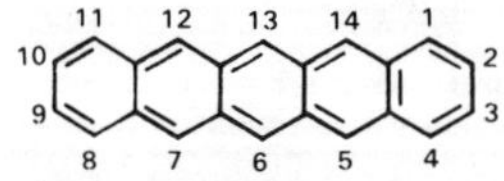	pentacene	pentacenyl-

[a]Listed in reverse order of preference
[b]*indicate exceptions to numbering and/or radical naming rules

TABLE 2

COMMON ABBREVIATED PREFIXES FOR FUSED RING SYSTEMS

Prefix	Parent Compound	Prefix	Parent Compound
Acenaphtho	Acenaphthylene	Naphtho	Naphthalene
Anthra	Anthracene	Perylo	Perylene
Benzo	Benzene	Phenanthro	Phenanthrene

peripheral sides of the base components a, b, c, etc., beginning with "a" for the side "1,2", "b" for the side "2,3" or "2,2a" and continuing until every side is lettered around the periphery (IUPAC Rule A21.5). An example of the application of these rules is shown in equation (1) for benzo[c]phenanthrene, a benzologue of phenanthrene:

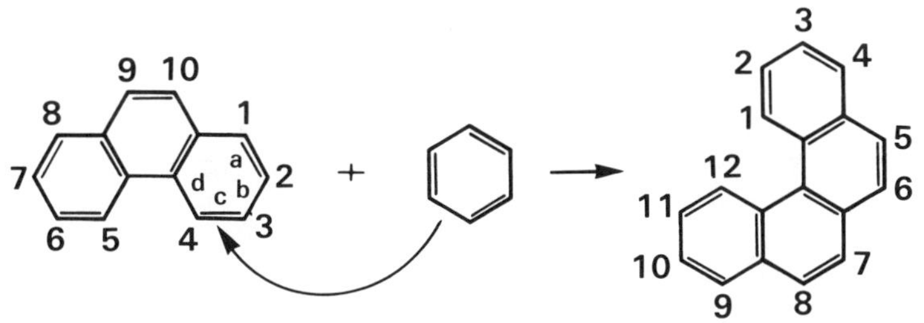

(1)

Phenanthrene **Benzo** **Benzo[*c*]phenanthrene**

Another PAH nomenclature rule which is particularly important for the naming of exocyclic N, O and S functional PAC, such as amino-PAH and hydroxy-PAH, involves radical names obtained from the trivial names. For radicals derived from PAH, the numbering of the hydrocarbon is retained, the point of attachment is given the lowest possible number, and the ending of the name is changed from "-ene" to "-enyl" (IUPAC Rules 24.1 and A24.2). Again, exceptions exist. The radical names corresponding to the trivially named hydrocarbons are listed in Table 1.

N-PAC NOMENCLATURE

Rules of nomenclature for the major N-PAC classes are governed by the extended Hantzsch-Widman System (IUPAC, Section B), fusion nomenclature or replacement nomenclature (IUPAC, Section B4) for the heterocyclic N-PAC, and substitutive nomenclature (IUPAC Section C) for the exocyclic N-PAC.

Endocyclic N-PAC

Heterocyclic nitrogen systems include fused-ring compounds containing atoms of two different elements as ring members; nitrogen and carbon. An extension of the Hantzsch-Widman System used for monocyclic compounds in conjunction with ortho/peri fusion nomenclature is one of two methods applicable to the nomenclature of the two classes of endocyclic nitrogen heterocyclic PAC. Table 3 provides a listing of trival names which are retained for the N-PAC with a single nitrogen heteroatom (IUPAC Rule B2.11). Unsaturated rings containing a nitrogen atom have designated suffixes: compounds with a nitrogen heteroatom in a five-membered ring end in "-ole"; likewise, nitrogen in a six-membered ring has a compound name ending in "-ine" (IUPAC Rule B1.1, Table II).

Fused-ring compounds containing a nitrogen atom are named according to the fusion principles reviewed in the PAH nomenclature review section (IUPAC Rule A21) except that the trivial names given in Table 3 are used; i.e., the base component should be a nitrogen heterocyclic system with as many rings as possible and occurring as high as possible in Table 3 (IUPAC Rule B3). The numbering of fused-ring N-PAC systems should provide the nitrogen atom with the lowest possible number, while maintaining the orientation rules outlined for the PAH. Exceptions to this rule are carbazole and acridine (see Table 3). A nitrogen atom common to two or more rings is numbered with the next whole number in sequence, in contrast to carbon atoms at fusion positions that are numbered by adding letters a, b, c, etc. to the previous whole number [IUPAC Rule B3.4(e)]. Lettering of peripheral sides for distinction of fusion isomers is also done by the PAH method. When the name of a component in a fusion name contains locants (numerals or letters) that do not also apply to the numbering of the fused system, these locants are placed in square brackets, as are the locants for fusion positions required by the PAH nomenclature (IUPAC Rule B3.1). For example, the structure obtained by fusing a benzene ring on the "h" side of quinoline shown in

TABLE 3

IUPAC STRUCTURES, NUMBERING, AND TRIVIAL NAMES FOR SELECTED N-PAC

IUPAC No.[a]	Structure	Trivial Name	Radical Name[b]
(12)	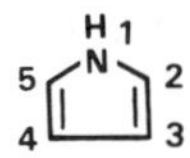	pyrrole or 1*H*-pyrrole[c]	pyrrolyl- or 1*H*-pyrrolyl-
(15)	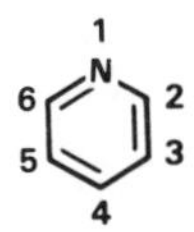	pyridine	pyridyl-*
(22)	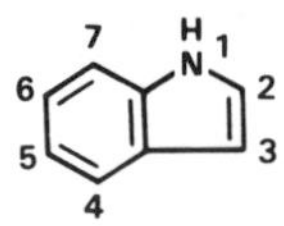	indole or 1*H*-indole[c]	indolyl- or 1*H*-indolyl-
(26)	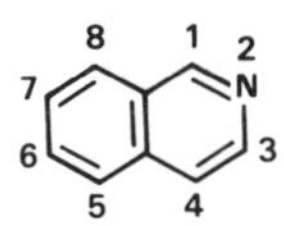	isoquinoline	isoquinolyl-*
(27)	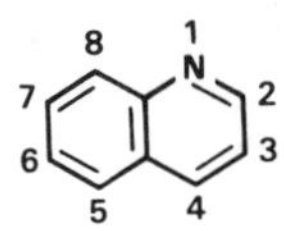	quinoline	quinolyl-*
(35)	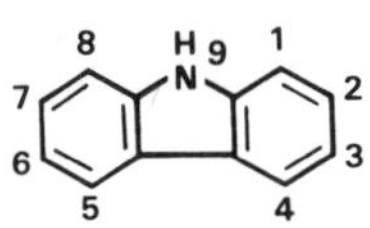	carbazole* or 9*H*-carbazole*[c]	carbazolyl- or 9*H*-carbazolyl-
(37)	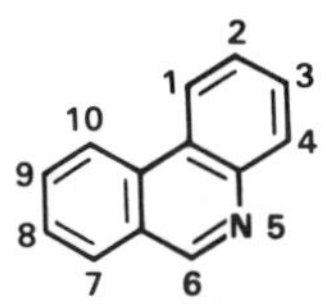	phenanthridine	phenanthridinyl-
(38)	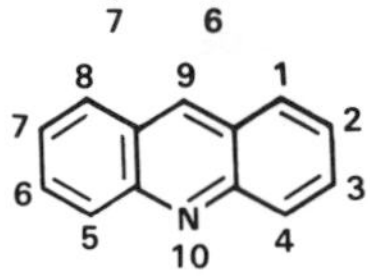	acridine*	acridinyl-

[a]Listed in reverse order of preference
[b]*Indicates exceptions to numbering and/or radical naming
[c]Currently used in Chemical Abstracts

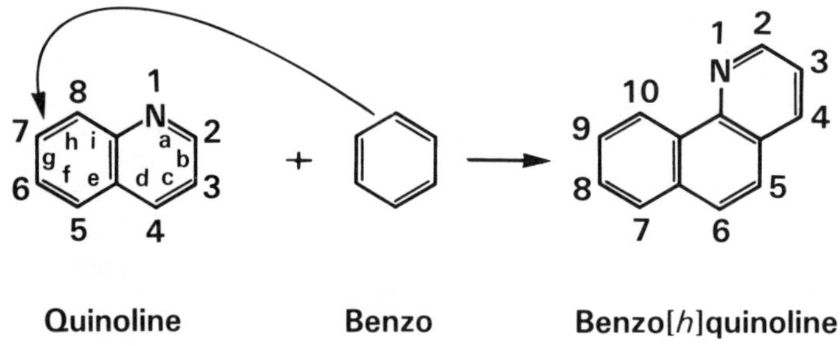

(2)

equation (2) has the systematic name of benzo[h]quinoline. The spatial orientation and numbering of this compound are in accordance with those rules reviewed for the PAH.

The vowel-elision rule (IUPAC A21.4 ** footnote) is frequently used in naming heterocyclic N-PAC. The final "o" or "a" of some fusion component prefixes is elided if the base component begins with a vowel. For example, benz(o)[h]isoquinoline becomes benz[h]isoquinoline and benz(o)[c]acridine is benz[c]acridine. Other N-PAC with a single nitrogen heteroatom and three or more fused-rings can be named by the application of these fundamental rules.

A few examples of naming endocyclic N-PAC are given in Figure 1, structures I-IV. Two additional points are noteworthy from the examples of Figure 1. First, as shown for structure III, the location of a hydrogen atom in a polycyclic system not completely unsaturated is identified by using a capital italic "H" in conjunction with the numerical locant (IUPAC Rules A21.6 and B5.12). Further hydrogenation is indicated by the prefixes "dihydro-", "tetrahydro-", etc. (IUPAC Rule B1.2). Secondly, from structure IV, the assignment of locants in naming complex fused-ring systems follows two additional rules: (a) locants are assigned the lowest possible numbering and lettering schemes; and (b) the attachment points numbering sequence for the substituting system(s) follows the direction of attachment as specified by the sequence of lettered locants on the base component system (IUPAC Rule A21.5).

Replacement Nomenclature

Nitrogen-containing fused ring systems may also be named by prefixing the name of the corresponding hydrocarbon with the term "aza-" (IUPAC Rule B1.1, Table 1), preceded by the locant of the heteroatom (IUPAC Rule B4.2). The structures,

I

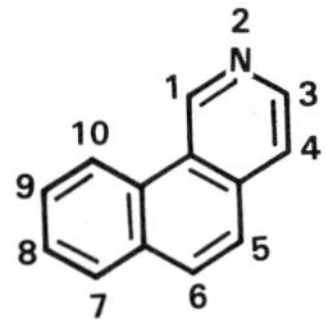

Benzo[*h*]isoquinoline
or
3-Azaphenanthrene

II

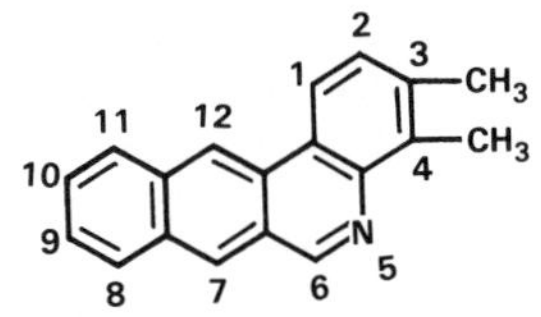

3,4-Dimethylbenzo[*j*]phenanthridine
or
3,4-Dimethyl-5-azabenz[*a*]anthracene

III

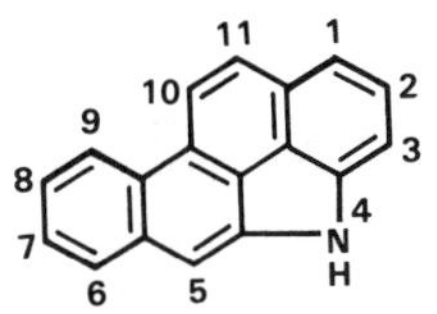

4*H*-Dibenzo(*b,def*]carbazole
or
4*H*-Azacyclopenta[*def*]chrysene

IV

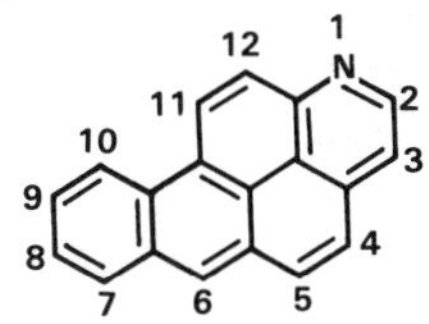

Anthra[2,1,9-*def*]quinoline
or
1-Azabenzo[*a*]pyrene

V

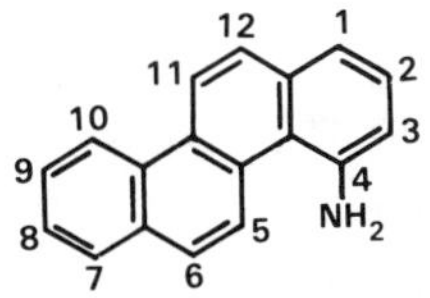

4-aminochrysene
4-chrysenylamine
4-chrysenamine

VI

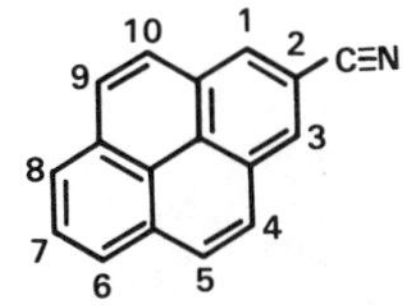

2-cyanopyrene
2-pyrenecarbonitrile
2-pyrenyl cyanide

VII

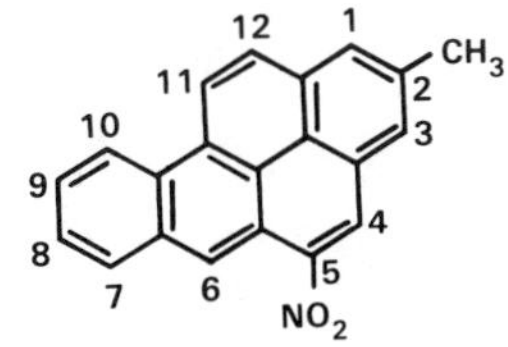

2-methyl-5-nitrobenzo[*a*]pyrene

FIGURE 1. Names of Selected N-PAC

numbering, and names of PAH in Table 1 can be used directly in naming endocyclic N-PAC. This method is known as replacement nomenclature, sometimes also referred to as "a" nomenclature. The numbering of the hydrocarbon is retained irrespective of the atom position. However, the lowest possible number consistent with the numbering of the PAH is assigned as the locant of the nitrogen heteroatom. In fusion names, the "aza" term always precedes the completed name of the parent hydrocarbon. Prefixes for ordinary substitution such as alkylation precede the "aza" term (IUPAC Rule B4.3). The use of this system is illustrated by the alternate names for structures I-IV in Figure 1.

Substitutive Nomenclature

The naming of exocyclic N-PAC, such as the amino-PAH, cyano-PAH, and nitro-PAH is achieved by use of substitutive nomenclature as contained in Section C of the IUPAC rules [1]. The cyano and amino functional groups may be cited as either suffixes or prefixes to the name of the parent compound (IUPAC Rule C10.2, Table III); for the IUPAC preferred order of precedence of function see Table 5. The prefixes for the -C≡N and $-NH_2$ functional groups of N-PAC are "cyano-" and "amino-", respectively, while the suffixes are "-carbonitrile" and "-amine", respectively. PAH with a nitro functional group are cited only with the nitro prefix in substitutive nomenclature (IUPAC Rule C10.1, Table I).

Primary monoamines, RNH_2, where R is a PAH are named by adding (a) the suffix "-amine" to the name of the PAH radical (Table 1), (b) the suffix "-amine" to the name of the parent compound with elision of the terminal "e", or (c) the prefix "amino-" to the name of the parent compound (IUPAC Rule C812). Use of locants with the lowest possible numbering are also used for isomer distinctions. Table 1 can be conveniently used for naming the amino-PAH, as well as the cyano-PAH and nitro-PAH. For example, phenanthrene with an amino functional group attached at position nine can be named 9-aminophenanthrene, 9-phenanthrylamine, or 9-phenanthrenamine.

Similar rules of nomenclature are used for naming the cyano functional PAH (IUPAC Rule C832). Monocyano PAC, R-C≡N, are named by adding (a) the suffix carbonitrile to the parent compound name (IUPAC Rule C832.2), (b) cyanide to the parent radical name (IUPAC Rule C832.4), or (c) the prefix cyano to the parent PAH compound (IUPAC Rule C832.5). "Carbonitrile" used as a suffix denotes the group -C≡N, including the carbon atom contained therein (IUPAC Rule C832.2). For the same example used for the amino functional phenanthrene, the corresponding cyano compound would be named 9-cyanophenanthrene, 9-phenanthrene-carbonitrile, or 9-phenanthryl cyanide.

PAH with an $-NO_2$ functional group are named only by means of a prefix "nitro-" (IUPAC Rule 852.1). Thus, for the example cited for the cyano- and amino substitution of phenanthrene in position nine, the only possible name for the analogous nitro- substituted compound would be 9-nitrophenanthrene. A few examples of naming exocyclic N-PAC by substitutive nomenclature are shown in Figure 1, structures V-VII.

O-PAC NOMENCLATURE

Endocyclic O-PAC

The naming of ring systems containing an oxygen atom is governed by the same basic principles as for the N-PAC outlined in the preceding pages. Oxygen heterocyclic compounds may be named by the extended Hantzsch-Widman System (IUPAC, Rule B1), or as *ortho/peri* fused ring systems incorporating the extended Hantzsch-Widman names and a select list of trivial names (IUPAC, Rule B3). Alternatively oxygen heterocyclic compounds may be named by the skeletal replacement method (IUPAC, Rule B4).

Table 4 provides a listing of trivial names which are retained for the ring systems containing a single oxygen heteroatom (IUPAC, Rule B2.11). These names also are used in forming fusion names (IUPAC, Rule B1). Other oxygen monocycles are named according to the extended Hantzsch-Widman System (IUPAC, Rule B1.1, Table II). In this system the prefix "oxa" with the appropriate numerical prefixes di, tri etc. is combined with the stems -irene, -ete, -ole, or -in, to denote unsaturated rings of 3, 4, 5 or 6 members, etc. In forming fusion names in the absence of other heteroatoms, the base component is selected to contain the greatest number of rings, and if there is still a choice, to contain the largest possible individual ring, etc. (IUPAC, Rule B3.1).

Replacement Nomenclature

Alternatively, fused O-PAC systems may be named using replacement nomenclature by prefixing the term "oxa", preceded by the locant of the atom it is replacing, to the name

TABLE 4

RETAINED TRIVIAL NAMES OF O-PAC USED AS A BASIS FOR FUSION NAMES

IUPAC No.	Parent Compound	Name	Radical Name
(5)	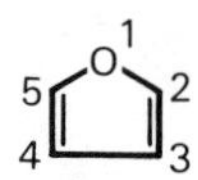	Furan	Furyl*
(6)	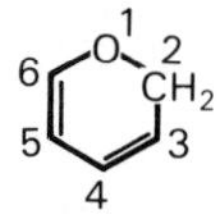	2*H*-Pyran	2*H*-Pyranyl
(7)	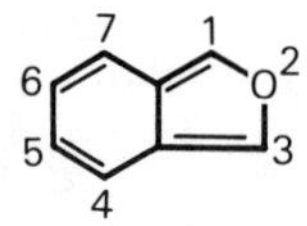	Isobenzofuran	Isobenzofuranyl
(8)	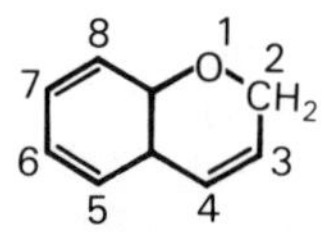	Chromene or 2*H*-1-Benzopyran[b]	Chromenyl or 2*H*-1-Benzopyranyl
(9)	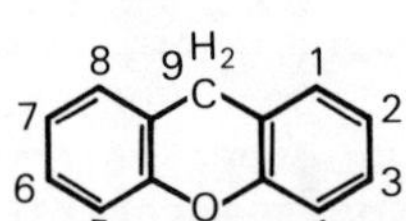	Xanthene* or 9*H*-Xanthene[b]	Xanthenyl or 9*H*-Xanthenyl

*Denotes exceptions to systematic numbering

*Indicates exceptions to numbering and/or radical naming rules
[b]As currently used in Chemical Abstracts

of the corresponding hydrocarbon. It is considered to be a skeletal modifying prefix as opposed to a substitutive prefix (substituting for hydrogen) and thus, is a "nondetachable prefix." It is cited directly in front of the name of the parent hydrocarbon, and it is not alphabetized with substituent prefixes or "radicals." Two examples of naming endocyclic O-PAC by fusion and replacement methods are given in Figure 2.

Indicated Hydrogen

Several naphtho[2,1-*b*]pyrans each containing the maximum number of noncumulative double bonds are possible. However, the name is made specific to the structure shown by indicating the position of the "extra" hydrogen in the structure. This is done by prefixing to the name, a locant and an italic capital "*H*" for this hydrogen atom. This technique is called "indicated hydrogen" (IUPAC, Rule A21.6). Similarly, the technique is applied to the corresponding replacement name 4*H*-1-oxaphenanthrene in Figure 2.

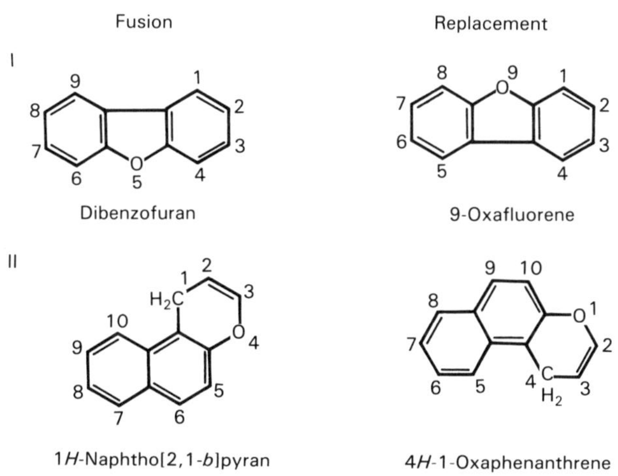

FIGURE 2. Comparison of Fusion and Replacement Methods

Exocyclic O-PAC

Substitution of hydrogen atoms of a parent hydride or PAH by other atoms or functional groups is denoted by prefixes or suffixes as was discussed in the N-PAC section and corresponding suffixes for some commonly occurring groups can be found in Table III of IUPAC Section C, and copied here for Table 5. These groups are listed in order of increasing priority for selection as the principal group. Preferably, the principal group is cited as a suffix (IUPAC, Rule C10.3). Thus, in Figure 3, the name fluoren-9-one is preferred to 9-oxofluorene.

In naming alcohols and phenols by substitutive nomenclature the suffix "-ol" or the prefix "hydroxy" denotes the substitution of hydrogen by an -OH group as for example 2-chrysenol or 2-hydroxychrysene, Figure 3. There is no substitutive suffix for denoting ethers, thus the ether group must be cited as a prefix as in 2-methoxychrysene, Figure 3.

I

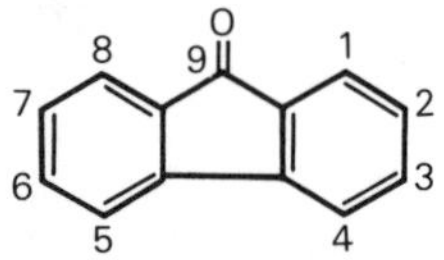

Fluoren-9-one
or
9-Oxofluorene

II

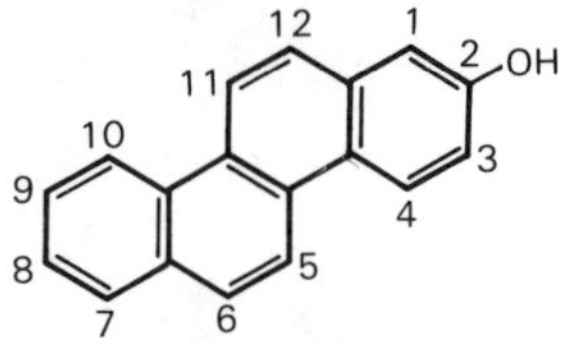

2-Chrysenol
or
2-Hydroxychrysene

III

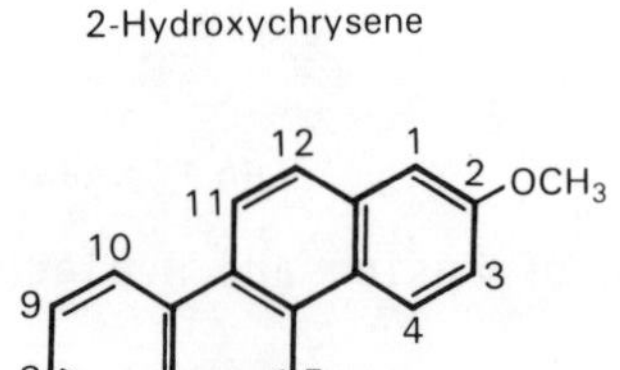

2-Methoxychrysene

FIGURE 3. Examples of Some Substituted O-PAC

TABLE 5

SUFFIXES AND PREFIXES AND THE ORDER OF PRECEDENCE OF FUNCTIONS

Class	Formula[a]	Prefix	Suffix	IUPAC Rule
Cations		-onio -onia	-onium	C-82 B-6
Carboxylic acid	-COOH -(C)OOH	Carboxy -	-carboxylic acid -oic acid	C-401
Sulfonic acid	$-SO_3H$	Sulfo	-sulfonic acid	C-641
Salts	-COOM -(C)OOM	- -	Metal...carboxylate Metal...oate	C-461
Esters	-COOR -(C)OOR	R-oxycarbonyl -	R...carboxylate R...oate	C-463
Acid halides	-CO-Halogen -(C)O-Halogen	Haloformyl -	-carbonyl halide -oyl halide	C-481
Amides	$-CO-NH_2$	Carbamoyl	-carboxamide	C-821
	$-(C)O-NH_2$	-	-amide	C-822
Amidines	$-C(=NH)-NH_2$ $-(C)(=NH)-NH_2$	Amidino -	-carboxamidine -amidine	C-951
Nitriles	$-C\equiv N$ $-(C)\equiv N$	Cyano -	-carbonitrile -nitrile	C-832
Aldehydes	-CHO -(C)HO	Formyl Oxo	-carbaldehyde -al	C-301 C-301
Ketones	$>(C)=O$	Oxo	-one	C-311
Alcohols	-OH	Hydroxy	-ol	C-201
Phenols	-OH	Hydroxy	-ol	C-202
Thiols	-SH	Mercapto	-thiol	C-511
Hydroperoxides	-O-OH	Hydroperoxy	-	C-218
Amines	$-NH_2$	Amino	-amine	C-812
Imines	=NH	Imino	-imine	C-815
Ethers	-OR	R-oxy	-	C-211
Sulfides	-SR	R-thio	-	C-514
Peroxides	-O-OR	R-dioxy	-	C-218

[a]Carbon atoms enclosed in parentheses are included in the name of the parent compound and not in the suffix or prefix (IUPAC, Rules C11.11, C11.31, etc.)

In naming ketones of unsaturated or aromatic systems the suffix "-one" denotes replacement of either $=CH_2$ or $\geqslant CH$ by $=CO$. If possible, it is preferable to replace a $=CH_2$ group by $=CO$ as in fluoren-9-one of Figure 3. When this is not possible, as in, for example, the 9-anthracenones, "added hydrogen" similar to "indicated hydrogen" is cited to denote one of several isomers that differ only with respect to the location of hydrogen atoms. "Added hydrogen," however, is cited in parentheses after the locant of the $=CO$ group, as shown in Figure 4 for 9(10_H_)-anthracenone (IUPAC, Rule C315). Alternatively a ketone may be denoted by the substitutive prefix "oxo" with the hydrogenation indicated by hydro prefixes; in this case the hydrogenation is considered to have occurred before the introduction of the keto group (IUPAC, C-316), as in 9,10-dihydro-9-oxoanthracene in Figure 4.

I

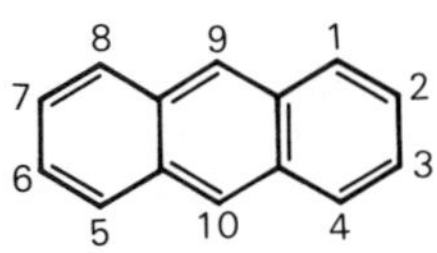

Anthracene

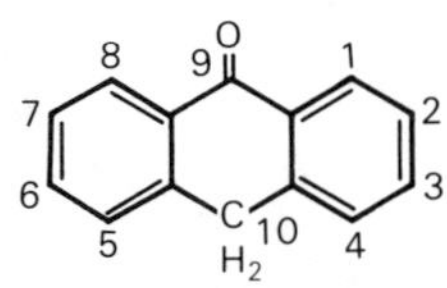

9(10*H*)-Anthracenone

II

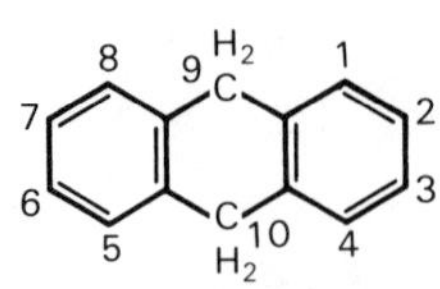

9,10-Dihydroanthracene

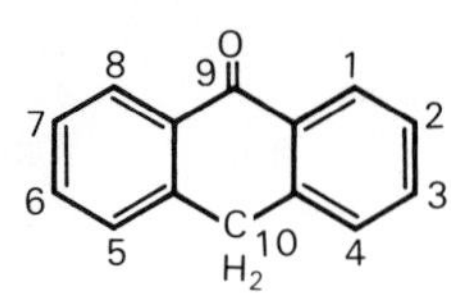

9,10-Dihydro-9-oxoanthracene

III

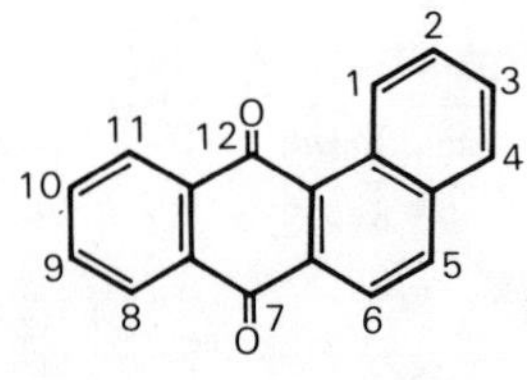

Benz[*a*]anthracene-7,12-quinone
or
Benz[*a*]anthracene-7,12-dione

FIGURE 4. Examples of Ketones and Quinones

Finally, diketones derived from aromatic compounds by conversion of two ⩾CH groups into =CO groups with any necessary rearrangement of double bonds to a quinoid structure are named by adding the suffix "quinone" or "dione" with locants to the name of the aromatic compound, Figure 4.

S-PAC NOMENCLATURE

Endocyclic S-PAC

Table 6 gives the structures and trivial names for the S-PAC which are retained for the compounds and used as a basis for fusion names (IUPAC Rule B2.11). Also given are the radical names which are formed according to IUPAC Rule B5. Fused-ring compounds containing a sulfur heteroatom are named according to the fusion principles presented previously in the PAH Nomenclature Review Section, except the base component should be one of the sulfur heterocyclic systems from Table 6 or a Hantzsch-Widman named monocycle.

TABLE 6

RETAINED TRIVIAL NAMES OF S-PAC USED AS A BASIS FOR FUSION NAMES

IUPAC No.	Parent Compound	Name	Radical Name[a]
(1)	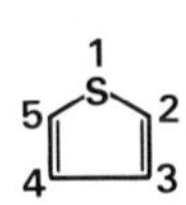	Thiophene	Thienyl-*
(4)	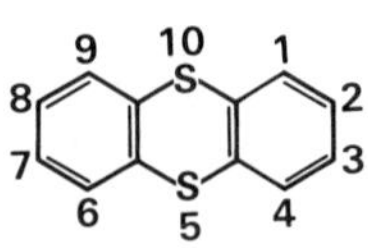	Thianthrene	Thianthrenyl-

[a]*Indicates exception to radical naming rules

Fusion isomers are distinguished by lettering the peripheral sides of the base sulfur-containing component. Thiophene has sides lettered "a" through "e" and thianthrene has sides lettered "a" through "n". To the letter as early in the alphabet as possible, denoting the side where fusion occurs, are prefixed, if necessary, the numbers of the positions of attachment of the attached PAH component. These numbers are chosen to be as low as is consistent with the numbering of the attached component and their order conforms to the direction of lettering of the base sulfur component. When two or more prefixes refer to equivalent positions so that there is a choice of letters (as is the case for the equivalent "b" and "d" sides of the thiophene ring), the prefixes are cited in alphabetical order and the location of the first cited prefix is indicated by the letter as early as possible in the alphabet. When two similar components are attached to the base component, numbering is distinguished with primes. The numbers and letters are enclosed in square brackets and placed immediately after the designation of the attached component (IUPAC Rule B21.5).

The complete fused system is oriented according to the principles given previously. After these principles, if there is a choice of orientations, the lowest number possible is given to the sulfur atom (IUPAC Rule B3.4). The properly oriented system is then numbered peripherally in a clockwise direction as previously described. The degree of hydrogenation is designated, if necessary, using the prefixes "dihydro", "tetrahydro", etc., followed by the name of the corresponding unreduced compound (IUPAC Rule A23.1), with location given by position number of the complete system. Substituted S-PAC are named with position number of the complete system and radical of the attached component as a prefix (IUPAC Rule A12). Some examples of proper S-PAC nomenclature and orientation are given in Figure 5. The interior letters and numbers given in Figure 5 refer to the base and attached components, respectively, for ease of interpretation of fused nomenclature principles. The exterior numbers refer to the complete fused system with proper orientation.

Fused S-PAC systems may also be named by prefixing the term "thia", preceded by its locants, to the name of the corresponding hydrocarbon (IUPAC Replacement Nomenclature Rule B4.2). The numbering of the corresponding hydrocarbon is retained, despite the position of the heteroatom. Dibenzothiophene (systematic name) illustrated in Figure 6 could then be called 9-thiafluorene (replacement name), with a

I

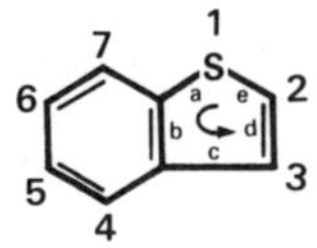

Benzo[*b*]thiophene

II

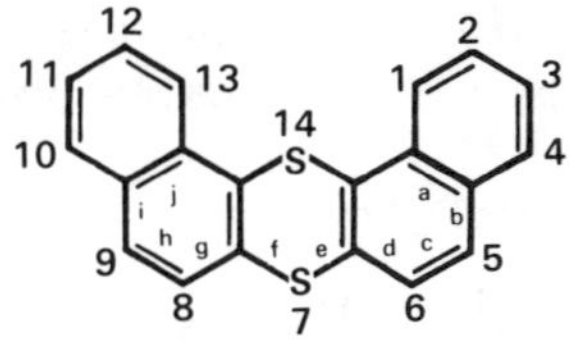

Dibenzo[*a,j*]thianthrene

III

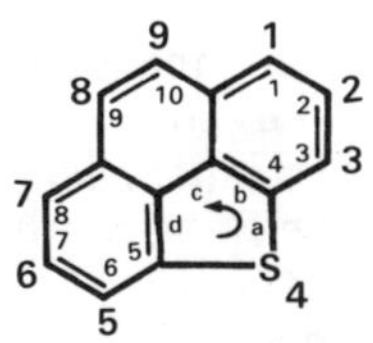

Phenanthro[*4,5-bcd*] thiophene

IV

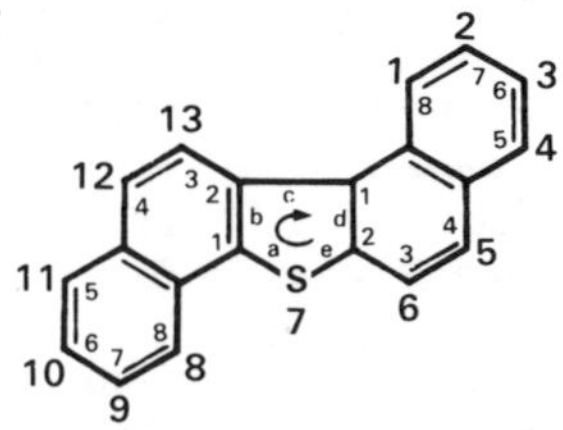

Dinaphtho[1,2-*b*:1′,2′-*d*]thiophene

V

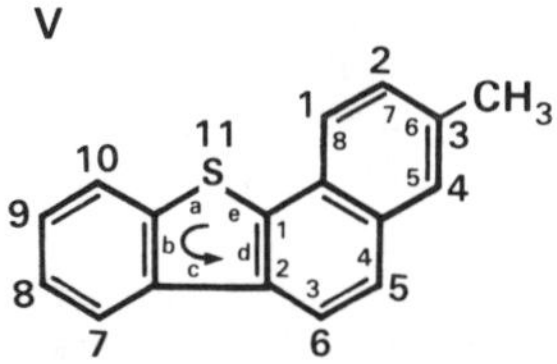

3-Methylbenzo[*b*]naphtho[2,1-*d*]thiophene

VI

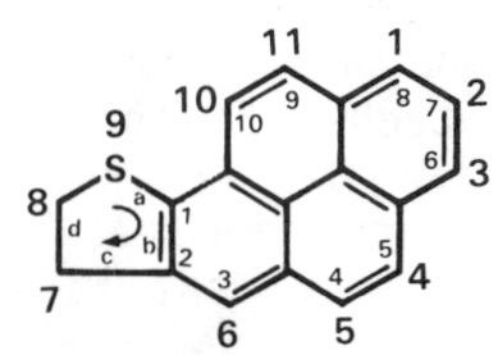

7,8-Dihydropyreno[1,2-*b*]thiophene

FIGURE 5. Examples of S-PAC nomenclature. Interior letters and numbers refer to the base and attached components. Exterior numbers refer to the complete fused system.

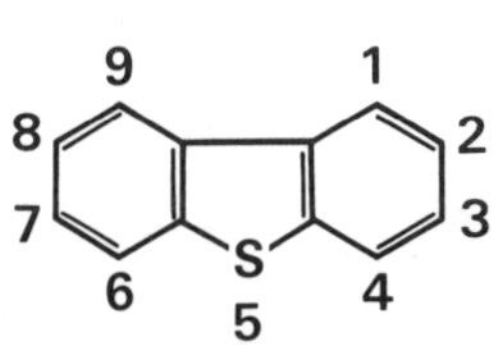

Dibenzothiophene

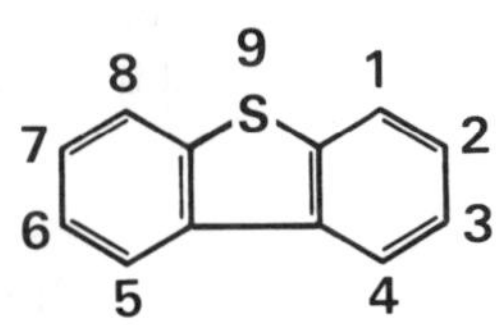

9-Thiafluorene

FIGURE 6. Dibenzothiophene or 9-Thiafluorene

different numbering system. For most applications, the systematic fusion names previously described are preferred.

Exocyclic S-PAC

In substitutive nomenclature, compounds containing -SH as the principal group directly attached to carbon are named by adding "-thiol" as a suffix or "mercapto-" as a prefix to the name of the parent compound (IUPAC Rule C511.1). The suffix is preferred. The position number of the carbon atom in the complete fused ring system is used for a locant. For example, an -SH group attached at the 2-position of the PAH compound phenanthrene would have the name 2-phenanthrenethiol or 2-mercaptophenanthrene. An -SH group attached to a fused ring system, such as the 4-position of benz[a]anthracene, would have the name benz[a]anthracene-4-thiol or 4-mercapto-benz[a]anthracene. Compounds containing -SR as the principal group directly attached to carbon are named by adding "alkylthio-" or "arylthio-" as a prefix to the name of the parent compound (IUPAC Rule C514.1), with position distinguished as was the case with -SH substitution. 1-(Methylthio)naphthalene and 4-(phenylthio)pyrene are examples for compounds of -SR substitution at the 1- and 4- positions of naphthalene and pyrene, respectively. In the case of -SR groups attached to the carbon atom, radiocofunctional nomenclature may also be used (IUPAC Rule C514.1). In this case, compounds are named as ethers except the term "sulfide" is used. Names for the same -SR examples previously given then become methyl 1-naphthyl sulfide and phenyl 4-pyrenyl sulfide, respectively.

ACKNOWLEDGEMENT

This work was supported in part by the U.S. Department of Energy Office of Health and Environmental Research under Contract DE-AC06-76RLO-1830 to the Pacific Northwest Laboratory.

REFERENCES

1. International Union of Pure and Applied Chemistry (1979): Nomenclature of Organic Chemistry: Sections A, B, C, D, E, F and H. Pergamon Press, Oxford, 559 pp.

2. Fletcher, J. H., Dermer, O. C., and Fox, R. B. (1974): Nomenclature of Organic Compounds. Advances in Chemistry Series 126, American Chemical Society, Washington, D.C., 337 p.
3. Jacob, J. and Grimmer, G. (1977): Nomenclature and structure of polycyclic aromatic hydrocarbons, Int. Agency Res. Cancer, 3(29):15-30.
4. Lee, M. L., Novotny, M. V., and Bartle, K. D. (1981): Analytical Chemistry of Polycyclic Aromatic Compounds. Academic Press, New York, pp. 1-3, 363-440.
5. Loening, K. L. and Merritt, J. E. (1983): Some aids for naming polycyclic aromatic hydrocarbons and their heterocyclic analogs. In: Polynuclear Aromatic Hydrocarbons: Formation, Metabolism, and Measurement, edited by M. Cooke and A. J. Dennis, pp. 819-843, Battelle Press, Columbus, Ohio.
6. Later, D. W. (1984): Nitrogen polycyclic aromatic compounds in coal-derived materials. In: Handbook of Polycyclic Aromatic Hydrocarbons, Vol. 2, edited by A. Bjorseth and T. Ramdahl, Marcel Dekker (in press).
7. Patterson, A. M., Capell, L. T., and Walker, D. F. (1960): The Ring Index. American Chemical Society, Washington, D.C., 1425 pp., and Supplements I (1963), II (1964) and III (1965).
8. Chemical Abstracts Service (1984): Ring Systems Handbook, American Chemical Society, Washington, D.C.

Chapter 3

NOMENCLATURE OF METABOLIC PRODUCTS OF PAH

KURT L. LOENING AND JOY E. MERRITT

INTRODUCTION

In previous years, we have discussed such topics as the nomenclature of polycyclic hydrocarbons and the nitrogen, oxygen, and sulfur heterocyclic analogs as well as the nomenclature of functional derivatives containing these elements (1,2,3). This year, we will focus on the topic of metabolic products of PAH. To keep the discussion simple, we will concentrate on the benz[a]anthracene derivatives shown in Figure 1 (4).

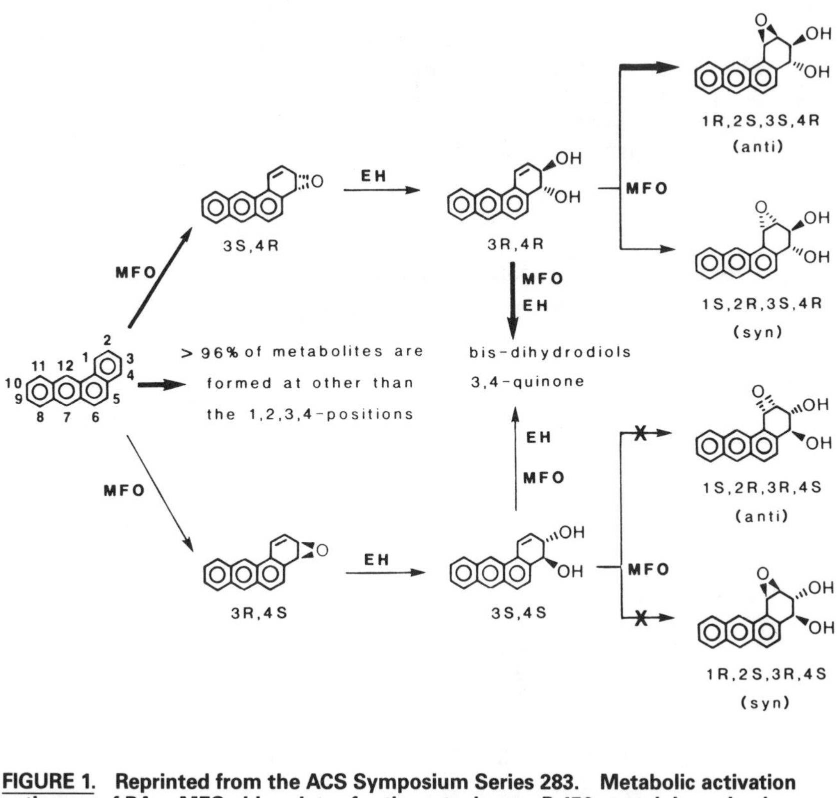

FIGURE 1. **Reprinted from the ACS Symposium Series 283. Metabolic activation pathways of BA. MFO abbreviates for the cytochrome P-450-containing mixed-function oxidases. The absolute configurations of the metabolites are as shown. (4)**

The nonstereochemical aspects of the nomenclature of these derivatives will be discussed first, followed by a discussion of the stereochemical descriptors.

The nomenclature recommendations of the International Union of Pure and Applied Chemistry (IUPAC) are codified in several books most readily identified by their colors. The Blue Book outlines the nomenclature of organic chemistry, while the Red Book codifies that of inorganic chemistry (5,6). A compendium of recommendations in the field of analytical chemistry is contained in the Orange Book, and the Green Book functions as a manual for physicochemical symbols, units, and terminology (7,8). The White Book, commonly referred to as The Compendium, contains the nomenclature recommendations of the IUPAC and the International Union of Biochemistry (IUB), Joint Commission on Biochemical Nomenclature (IUPAC-IUB/JCBN) (9). And finally, the book "Enzyme Nomenclature" is offered by the IUB (10).

More than one name may be formed for a compound in accordance with the IUPAC recommendations as is recognized by the Blue Book, which codifies existing practices. This book documents three types of nomenclature systems: substitutive, additive, and fusion nomenclature.

SUBSTITUTIVE NOMENCLATURE

With the substitutive method, characteristic groups or atoms are identified and cited either as prefixes or as a suffix depending on their rank in an established order of precedence of functions (see Table 1).

TABLE 1

SUFFIXES AND PREFIXES OF CHARACTERISTIC GROUPS IN THE ORDER OF DECREASING PRIORITY FOR CITATION AS PRINCIPAL GROUP

Class	Formula[a]	Prefix	Suffix
Cations		-onio -onia	-onium
Carboxylic acid	-COOH	Carboxy	-carboxylic acid
	-(C)OOH	-	-oic acid
Sulfonic acid	$-SO_3H$	Sulfo	-sulfonic acid
Salts	-COOM	-	Metal...carboxylate
	-(C)OOM	-	Metal...oate
Esters	-COOR	R-oxycarbonyl	R...carboxylate
	-(C)OOR	-	R...oate
Acid halides	-CO-Halogen	Haloformyl	-carbonyl halide
	-(C)O-Halogen	-	-oyl halide
Amides	$-CO-NH_2$	Carbamoyl	-carboxamide
	$-(C)O-NH_2$	-	-amide
Amidines	$-C(=NH)-NH_2$	Amidino	-carboxamidine
	$-(C)(=NH)-NH_2$	-	-amidine
Nitriles	-C=N	Cyano	-carbonitrile
	-(C)=N	-	-nitrile
Aldehydes	-CHO	Formyl	-carbaldehyde
	-(C)HO	Oxo	-al
Ketones	>(C)=O	Oxo	-one
Alcohols	-OH	Hydroxy	-ol
Phenols	-OH	Hydroxy	-ol
Thiols	-SH	Mercapto	-thiol
Hydroperoxides	-O-OH	Hydroperoxy	-
Amines	$-NH_2$	Amino	-amine
Imines	=NH	Imino	-imine
Ethers	-OR	R-oxy	-
Sulfides	-SR	R-thio	-
Peroxides	-O-OR	R-dioxy	-

[a]Carbon atoms enclosed in parentheses are included in the name of the parent compound and not in the suffix or prefix (IUPAC, Rules C11.11, C11.31, etc.)

When cited as prefixes, these groups are arranged in alphabetical order among themselves. Hydro prefixes are cited either within this order, or they may be cited directly in front of the name of the parent compound in accordance with IUPAC Rule C-16.33:

> "Hydro prefixes may be treated as detachable and arranged in alphabetical order among other detachable prefixes or they may be treated as non-detachable from the name of the parent compound or radical."

Example:

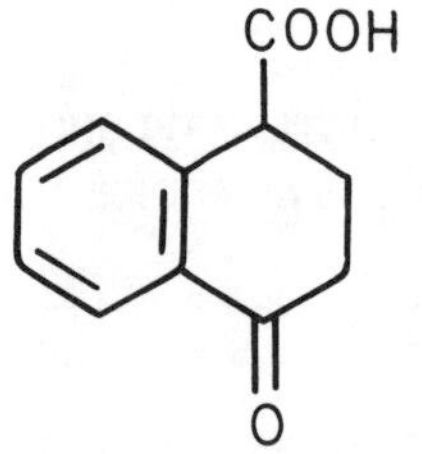

Detachable: 1,2,3,4-Tetrahydro-4-oxo-1-naphthoic acid

Non-detachable: 4-Oxo-1,2,3,4-tetrahydro-1-naphthoic acid

When a characteristic group is cited as a prefix, it is formally considered to substitute for one or more hydrogen atoms of a conceptual parent structure. This is in contrast to the additive method of nomenclature in which atoms are simply added to the conceptual parent structure, leaving the total number of hydrogen atoms in the compound unchanged.

We will use the epoxide or oxide derivatives of benz[a]anthracene to illustrate the differences between the substitutive and additive nomenclature methods. With the substitutive method, IUPAC Rule C-212.2 states:

> "An oxygen atom directly attached to two carbon atoms already forming part of a ring system or to two carbon atoms of a chain may be indicated by the prefix 'epoxy-', particularly when it is desired to preserve the name of a specific complex structure, as, for example, steroids or carotenoids."

Example:

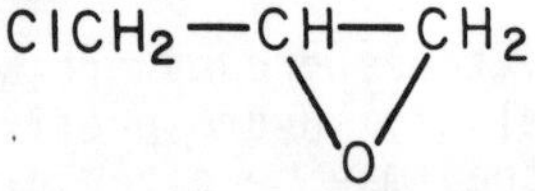

Substitutive name: 1-Chloro-2,3-epoxypropane

Thus, metabolic products of benz[a]anthracene are named by the substitutive nomenclature method as shown in Figure 2.

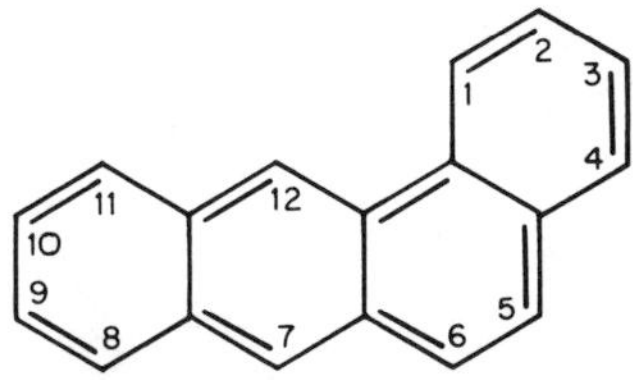

Parent: Benz[a]anthracene

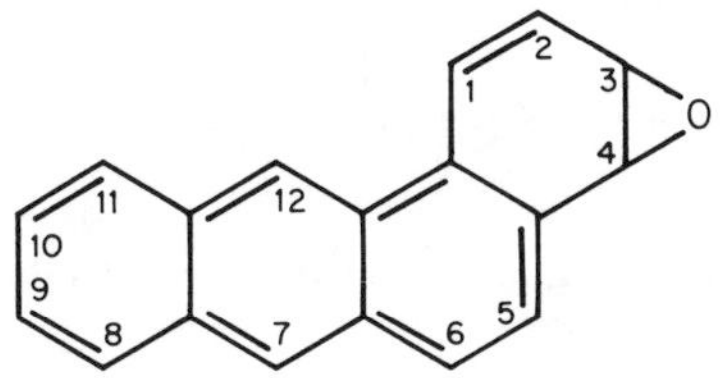

3, 4-Epoxy-3, 4-dihydrobenz[a]anthracene

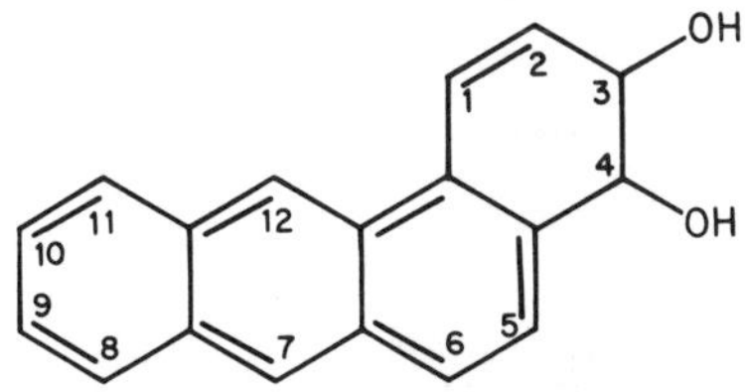

3, 4-Dihydrobenz[a]anthracene-3, 4-diol

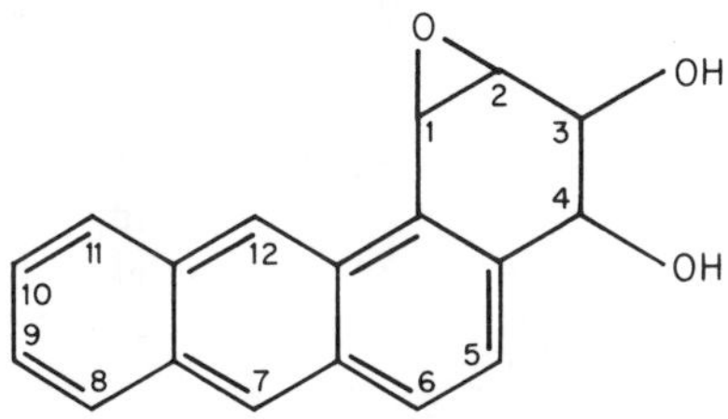

1, 2-Epoxy-1, 2, 3, 4-tetrahydrobenz[a]anthracene-3, 4-diol

FIGURE 2. Substitutive nomenclature.

In addition to placement of the hydro prefixes in a name, other variations within a name are permissible. For example, the principal characteristic group may be overstepped and cited as a prefix in a name in order to emphasize the ring itself. Some substitutive name variations are illustrated in Figure 3.

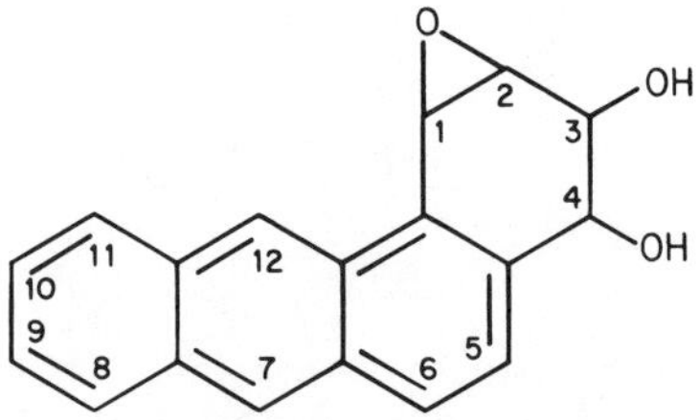

PREFERRED: 1, 2-Epoxy-1, 2, 3, 4-tetrahydrobenz[a]anthracene-3, 4-diol

OTHER: 1, 2-Epoxy-1, 2, 3, 4-tetrahydro-3, 4-dihydroxybenz[a]anthracene
1, 2-Epoxy-3, 4-dihydroxy-1, 2, 3, 4-tetrahydrobenz[a]anthracene

WRONG: 1, 2, 3, 4-Tetrahydro-3, 4-dihydroxy-1, 2-epoxybenz[a]anthracene

FIGURE 3. Substitutive name variations.

ADDITIVE NOMENCLATURE

Another type of nomenclature system, the additive method, may be used to name these PAH metabolic products. In contrast to the substitutive method, IUPAC Rule C-33.1 states

> "In additive nomenclature, except for hydro prefixes, the name of the added atoms is placed, in its ion form, after the name of the parent compound. This method of nomenclature is in general discouraged, although it is useful in a few specialized cases, particularly when the stereochemistry of addition is unknown or when the systematic name of the adduct might obscure the relation to the parent compound;"

Example:

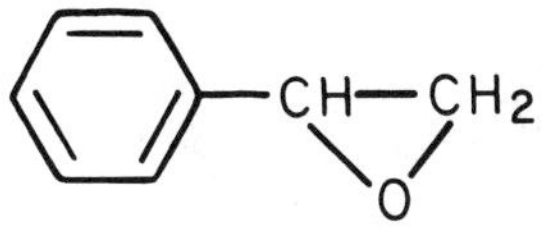

Additive name: Styrene oxide

or

Styrene epoxide

Substitutive names: 2-Phenyloxirane

or

(1,2-Epoxyethyl)benzene

Thus, the epoxides of benz[a]anthracene are named by additive nomenclature as shown in Figure 4.

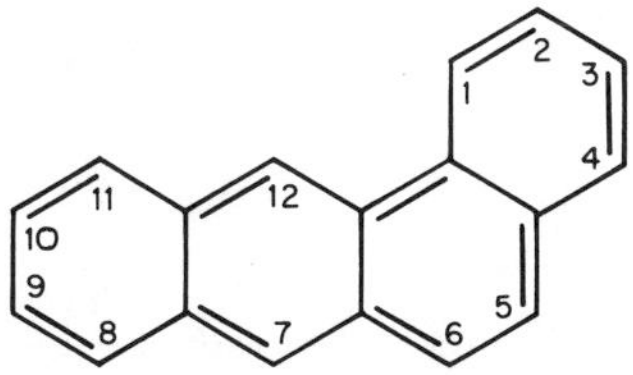

Parent: Benz[a]anthracene

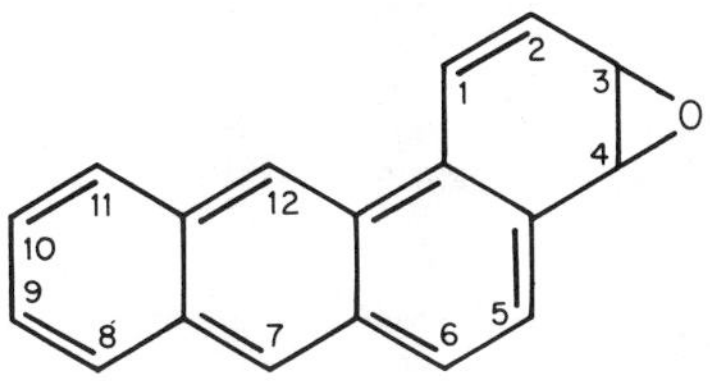

Benz[a]anthracene 3, 4-oxide

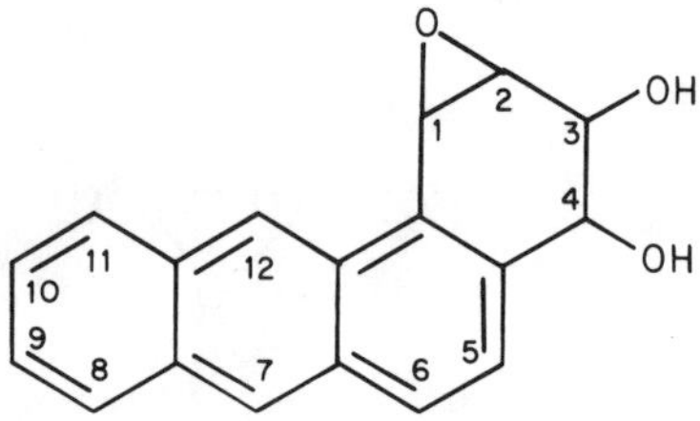

3, 4-Dihydrobenz[a]anthracene-3, 4-diol 1, 2-oxide

FIGURE 4. Additive nomenclature.

With this method, the term epoxide is a permissible variation of the term oxide. Some additive name variations are illustrated in Figure 5.

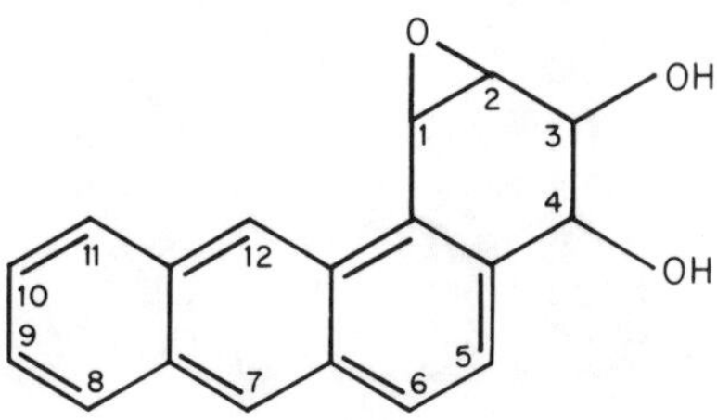

PREFERRED: 3, 4-Dihydrobenz[a]anthracene-3, 4-diol 1, 2-oxide

OTHER: 3, 4-Dihydrobenz[a]anthracene-3, 4-diol 1, 2-epoxide
3, 4-Dihydro-3, 4-dihydroxybenz[a]anthracene 1, 2-epoxide

WRONG STOICHIOMETRY: 1, 2, 3, 4-Tetrahydrobenz[a]anthracene-3, 4-diol 1, 2-epoxide

CLASS NAMES: Dihydrodiol oxide; Dihydrodiol epoxide;
3, 4-Dihydrodiol 1, 2-epoxide

FIGURE 5. Additive name variations.

FUSION NOMENCLATURE

Finally, a third method may be used to name benz[a]anthracene epoxide derivatives: the fusion method. This type of nomenclature is documented in IUPAC Rule A-21.3 for hydrocarbon rings and in IUPAC Rule B-3 for heterocyclic rings (5). This method was also extensively discussed and presented in the proceedings of previous PAH symposia (1,2,3). It is important to recall that in fusion nomenclature, rings containing the maximum number of non-cumulative double bonds are fused, and the completed system is then oriented and renumbered. Figure 6 illustrates the naming of benz[a]anthracene epoxide derivatives by fusion nomenclature.

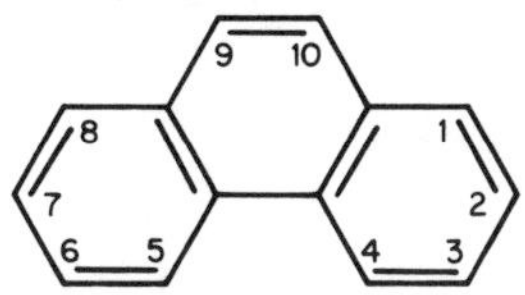

Phenanthrene

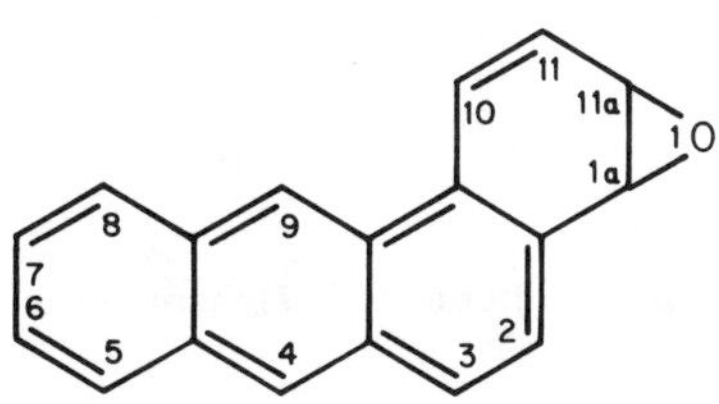

1a, 11a-Dihydrobenzo[6, 7]phenanthro[1, 2-b]oxirene

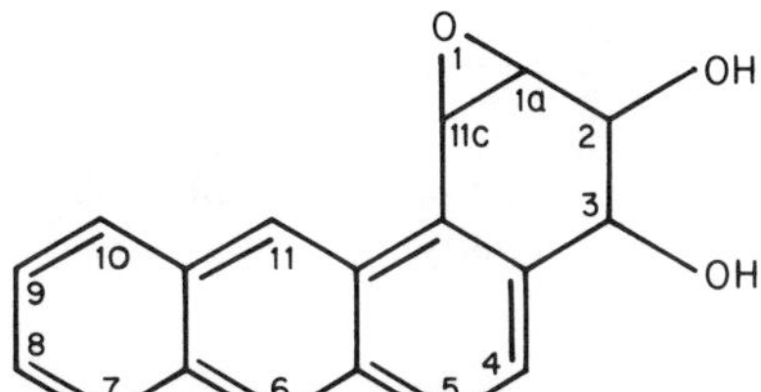

1a, 2, 3, 11c-Tetrahydrobenzo[6, 7]phenanthro [3, 4-b]oxirene-2, 3-diol

FIGURE 6. Fusion nomenclature.

For ease of comparison, three types of names derived by substitutive, additive, and fusion methods respectively are illustrated for a single compound in Figure 7.

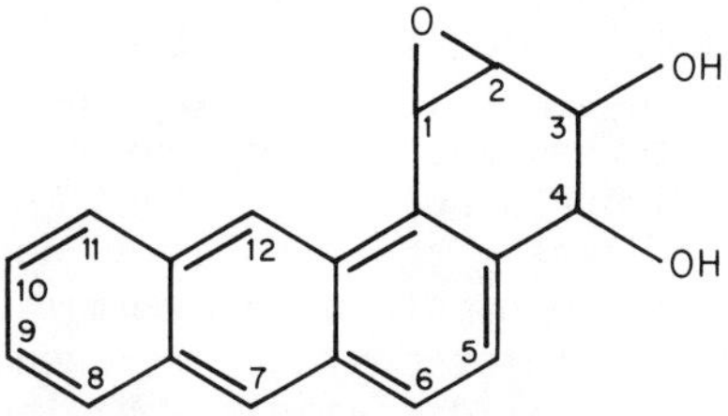

SUBSTITUTIVE: 1, 2-Epoxy-1, 2, 3, 4-tetrahydrobenz[a]anthracene-3, 4-diol

ADDITIVE: 3, 4-Dihydrobenz[a]anthracene-3, 4-diol 1, 2-oxide

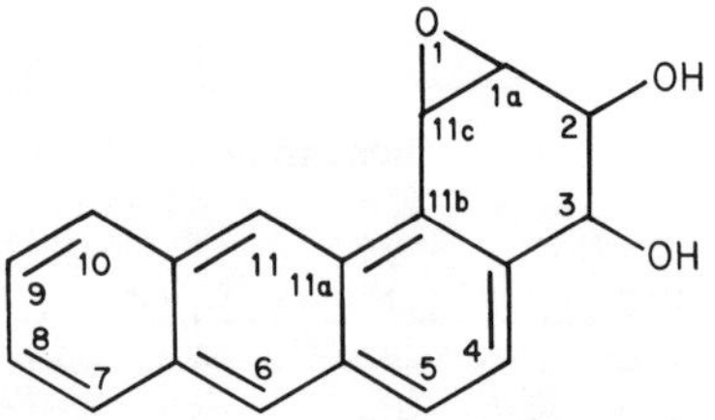

FUSION: 1a, 2, 3, 11c-Tetrahydrobenzo[6, 7]phenanthro[3, 4-b]oxirene-2, 3-diol

FIGURE 7. IUPAC preferred nomenclature.

Nomenclature errors occur frequently in the literature, especially in titles and text where class names are used. The stoichiometry is frequently confused between the substitutive and additive names. While it is correct to call the compounds in Figure 8 epoxytetrahydronaphthalenes, it is incorrect to call them naphthalene tetrahydro epoxides (11).

SUBSTITUTIVE CLASS NAME: Epoxytetrahydronaphthalenes

ADDITIVE CLASS NAME: Dihydronaphthalene oxides or Dihydronaphthalene epoxides

FUSION CLASS NAME: Tetrahydronaphthoxirenes

FIGURE 8. Class names.

Other errors commonly encountered include disregard for the alphabetical order. Even in combination with abbreviations, such as DMBA, alphabetical order can be maintained. Use of the name 3,4-Dihydroxy-1,2-epoxy-1,2,3,4-tetrahydro-DMBA is incorrect (12), but 1,2-Epoxy-1,2,3,4-tetrahydro-3,4-dihydroxy-DMBA is better. Still another error occurs when authors cite a characteristic group both as a prefix and as a suffix, and when they cite the hydro groups with the suffix as in 1-Hydroxycholanthrene-9,10-dihydrodiol-7,8-epoxide (13). The correct name for this compound is 9,10-Dihydrocholanthrene-1,9,10-triol 7,8-epoxide, or 9,10-Dihydro-1,9,10-trihydroxycholanthrene 7,8-epoxide. Hyphenation also causes some problems in the literature. For example, epoxide should be separated by a space from the name it modifies.

STEREOCHEMISTRY

IUPAC has carefully developed recommendations for the definitions and use of stereochemical descriptors. These are documented as Section E in the IUPAC Blue Book (5). To illustrate how the stereochemical configurations of the PAH are described in the IUPAC system, we shall continue using the example of the benz[*a*]anthracenes. The absolute configurations of four 3,4-dihydrodiol 1,2-oxides of benz[*a*]anthracene have been reported (see Figure 1). Our discussion addresses both relative and absolute stereodescriptors.

Syn and Anti

The terms *syn* and *anti* are used often in the PAH field to describe the relative configurations of dihydrodiol epoxides (14); however, these terms have no official IUPAC definition at present. These descriptors have been used in the literature to describe the geometry of oximes, but IUPAC recommends their abandonment for such compounds (5, Rule E-2.2.2). Their use by Chemical Abstracts Service (CAS) is restricted to describing only the senior substituent on the Z bridge of a bicyclo[X.Y.Z]ane (see Figure 9) (15).

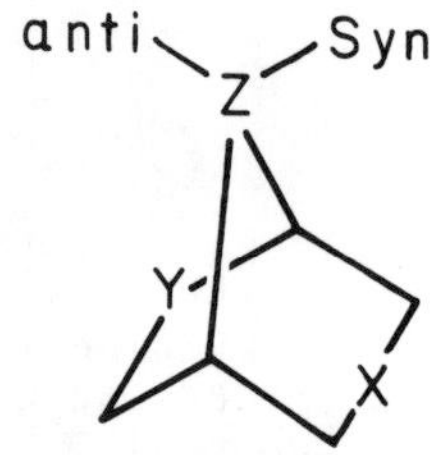

Bicyclo[X.Y.Z]ane

FIGURE 9. Use of *syn* and *anti* in *Chemical Abstracts*.

PAH chemists' use of *syn* and *anti* is unique to their field. Therefore, these terms may not be adequate for unambiguous communication to non-PAH chemists.

The Cahn-Ingold-Prelog Sequence Rule Procedure

If *syn* and *anti* cannot be used for interdisciplinary communication, what should the PAH chemist use in their place? Consider the absolute stereodescriptors *R* and *S* for describing absolute configurations (5, IUPAC Section E, 16). Assignment of the absolute terms *R* and *S* depends upon the priority ranking of atoms or groups attached to the stereochemical element whose chirality ("handedness") is to be determined by the Cahn-Ingold-Prelog Sequence Rule. This ranking depends first upon the descending order of atomic number of the atoms directly attached to the chiral center; thus, for bromochlorofluoroiodomethane the order is I, Br, Cl, F. In Figure 10 these atoms are represented by *a*, *b*, *c*, and *d*, respectively.

FIGURE 10. Assignment of the stereodescriptors *R* and *S*. The symbol *R* is assigned to a clockwise (right-handed) sequence of a, b, c, while *S* denotes a counterclockwise (anticlockwise) sequence.

The least preferred atom or group, d, is represented by a dotted line to indicate it is to be considered to lie below the plane of the paper, while a, b, and c are to be imagined to project toward the viewer at an angle. The analogy of an automobile steering wheel with three radial bars is useful to consider when visualizing this concept. The descriptor R is assigned to a clockwise (right-handed) sequence of a, b, c, while S denotes a counterclockwise (anticlockwise) sequence. This system is extremely comprehensive and can be used to describe almost all chiral centers. However R and S describe absolute configurations only. Syn and anti are relative descriptors, and therefore R and S cannot be used in their place.

The IUPAC Stereodescriptors R* and S*

To bridge this gap in the use of the Sequence Rule procedure, the IUPAC rules introduce the use of two new symbols, R* and S* for describing relative configurations (see Figure 11).

(1*R, 3*S**)-1-Bromo-3-chlorocyclohexane**

FIGURE 11. Assignment of stereodescriptors *R and *S** to denote relative configurations.**

These symbols are assigned by the same sequence procedure, but the center of chirality with the lowest locant is arbitrarily assigned the chirality *R*. These symbols are not part of the original paper by Cahn, Ingold, and Prelog on the "Specification of Molecular Chirality" in 1966. Nevertheless, they are a very important extension of the application of the Sequence Rule procedure. Like *R* and *S* the scope of *R** and *S** is very comprehensive, and thus they are excellent for use in interdisciplinary communication. Therefore, we recommend these IUPAC symbols be used in place of *syn* and *anti* for the PAH dihydrodiol oxides. Names that denote the relative configuration of a dihydrobenz[*a*]anthracenediol oxide derivative using *R** and *S** are shown in Figure 12.

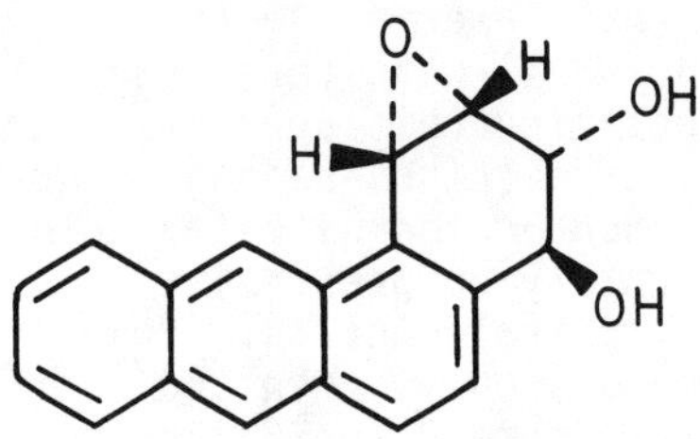

or

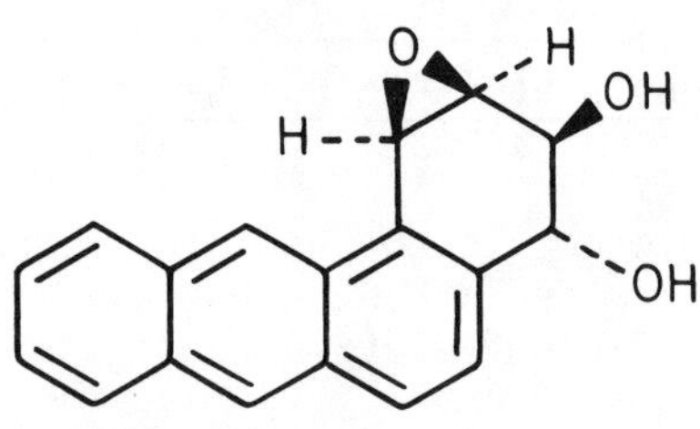

SUBSTITUTIVE: (1*R, 2*S**, 3*S**, 4*R**)-1, 2-Epoxy-1, 2, 3, 4-tetrahydrobenz[*a*]anthracene-3, 4-diol**

ADDITIVE: (1*R, 2*S**, 3*S**, 4*R**)-3, 4-Dihydrobenz[*a*]anthracene-3, 4-diol 1, 2-oxide**

FUSION: (1a*R, 2*R**, 3*S**, 11c*S**)-1a, 2, 3, 11c-Tetrahydrobenzo[6, 7]phenanthro[3, 4-*b*]oxirene-2, 3-diol**

FIGURE 12. Names that denote the relative configuration using *R and *S**.**

The R*/S* symbols may be cited together as a group in front of the entire name or placed within the name nearest the part to which they refer. These descriptors may also be used with either the substitutive or additive type names. However if used in the name in which the oxirene ring is expressed by fusion, the locants will change.

Racemates

Racemates are described either by placing the symbol (±) before the R*,S* relative stereodescription, or by the pairs RS or SR according to whether the center is R or S when the first is R (see Figure 13).

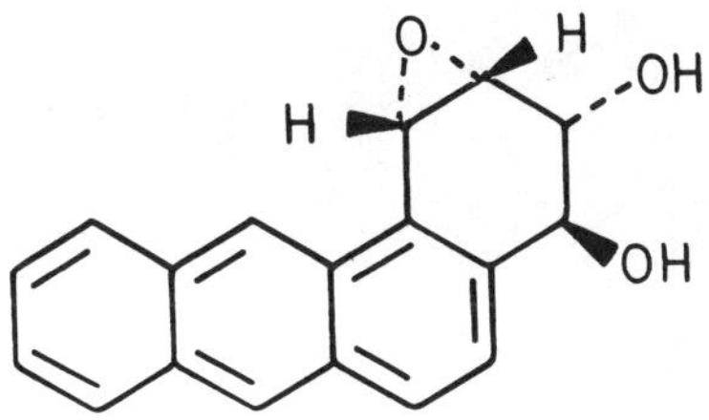

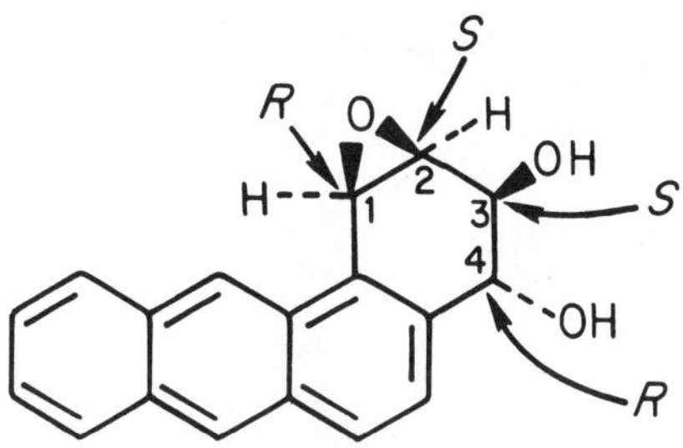

(±)-(1*R*, 2*S**, 3*S**, 4*R**)-**
or
(1*RS*, 2*SR*, 3*SR*, 4*RS*)-

FIGURE 13. IUPAC Stereodescription of a racemate.

The Absolute Stereodescriptors R and S

The absolute R and S stereodescriptors for the (+)-syn, (-)-syn, and similarly for the (+)- and the (-)-anti isomers are shown in Figure 14.

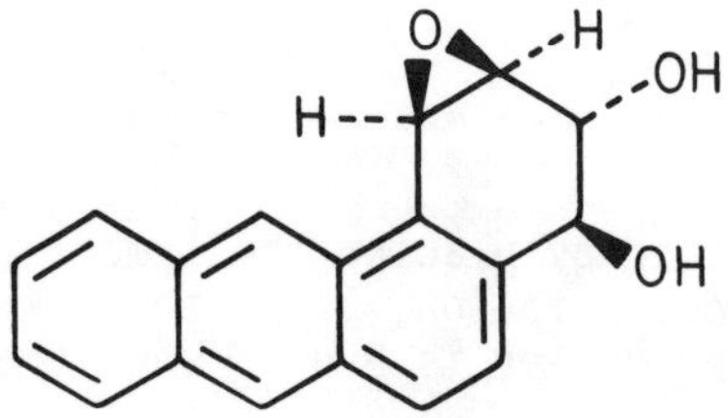

IUPAC: (+)-(1*R*, 2*S*, 3*R*, 4*S*)-
CAS: [1a*S*-(1aα, 2α, 3β, 11cα)]-
Other: (+)-syn-

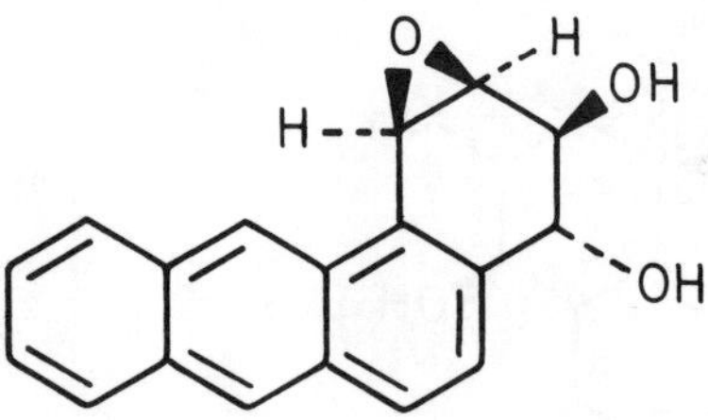

IUPAC: (+)-(1*R*, 2*S*, 3*S*, 4*R*)-
CAS: [1a*S*-(1aα, 2β, 3α, 11cα)]-
Other: (+)-anti-

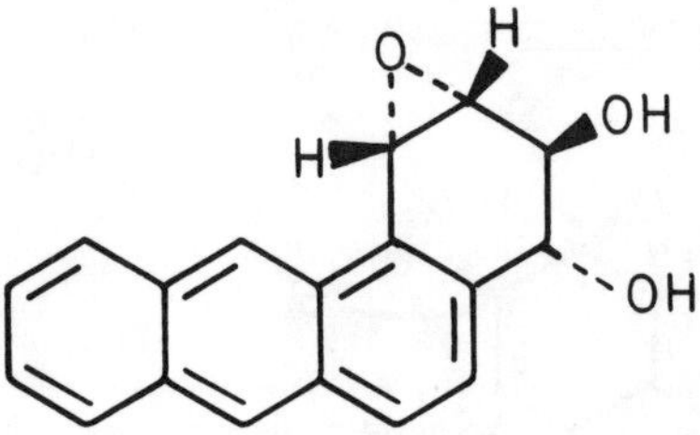

IUPAC: (−)-(1*S*, 2*R*, 3*S*, 4*R*)-
CAS: [1a*R*-(1aα, 2α, 3β, 11cα)]-
Other: (−)-syn-

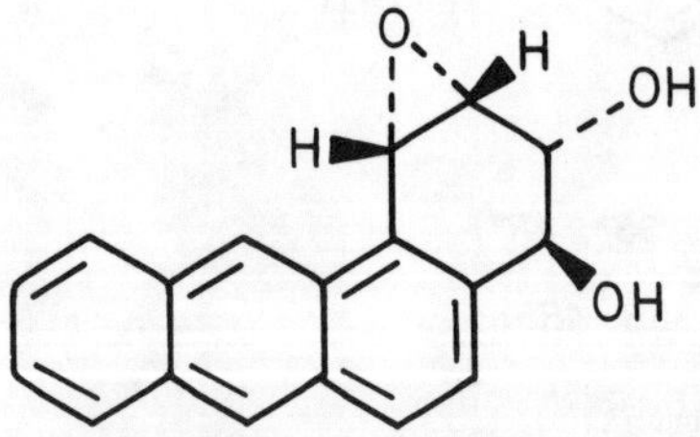

IUPAC: (−)-(1*S*, 2*R*, 3*R*, 4*S*)-
CAS: [1a*R*-(1aα, 2β, 3α, 11cα)]-
Other: (−)-anti-

FIGURE 14. **Stereodescriptors for the absolute configurations.**

In the actual assignment of R and S to these compounds it is easy to make an incorrect assignment at position 2 of 1,2-epoxy-1,2,3,4-tetrahydrobenz[a]anthracene-3,4-diol. This position is analogous to position 9 of 9,10- epoxy-7,8,9,10-tetrahydrobenzo[a]pyrene-7,8-diol (see Figure 15).

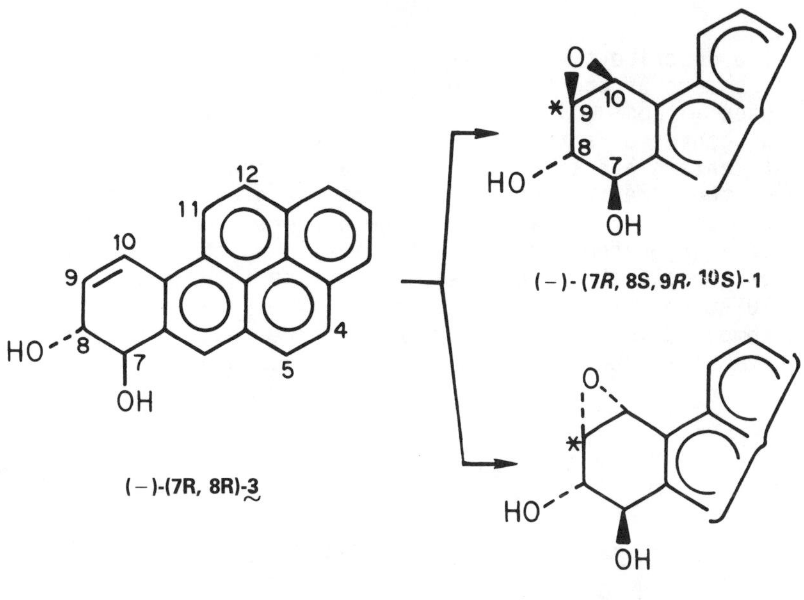

(–)-(7R, 8R)-3

(–)-(7*R*, 8S, 9*R*, 10S)-1

(+)-7*R*, 8S, 9S, 10R)-2

FIGURE 15. Metabolic conversion of (–)-(7R, 8R)-dihydrodiol ((–)-3) into the diastereomeric 7, 8-diol 9, 10-epoxides ((–)-1 and (+)-2).
Reprinted from the *J. Am. Chem. Soc.* 1977, 99(7) 2358.

If in the second level of exploration, one is not careful to explore the connections of the preferred pathway first before proceeding to explore the less preferred pathway, an incorrect assignment of R or S will result (16, 17).

CAS Stereodescriptors

The CAS system for describing stereogenic compounds is documented in Appendix IV of the Chemical Abstracts Index Guide (15). Stereodescriptors for systematically named compounds are covered in paragraph 203 of Appendix IV. This paragraph is also coincidentally a part of a section labelled "Section E," but should not be confused with Section E of the IUPAC Blue Book.

CAS Relative Stereodescriptors and Description of Racemates

To describe the relative configurations of the *syn* and *anti* isomers of the "dihydrodiol epoxide" of benz[*a*]anthracene CAS uses α and β in a relative sense. The first cited chiral center of a ring is arbitrarily designated α . Subsequent centers are cited as α or β according to whether they are on the same or opposite sides of the plane of the ring. It should be noted that this usage differs from that for cyclic stereoparents such as steroids in which " α " means "below the plane" and denotes absolute configuration. The CAS and IUPAC relative configurations of the *syn* and *anti* isomers are compared in Figure 16. Racemates are described by adding the symbol (±) to the relative stereodescriptor.

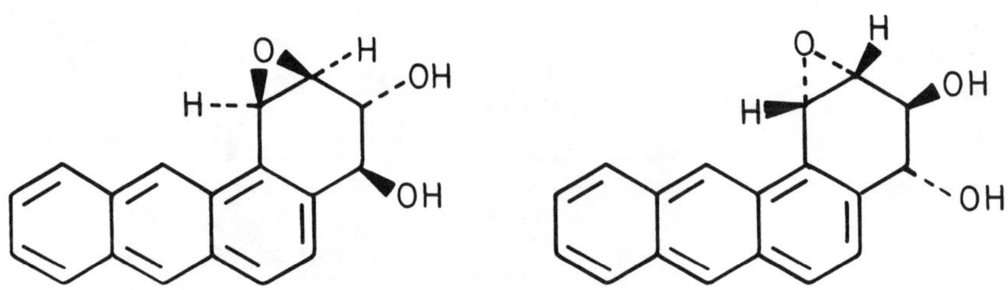

IUPAC: (1*R, 2*S**, 3*R**, 4*S**)-**
CAS: (1aα, 2α, 3β, 11cα)-
Other: *syn*-

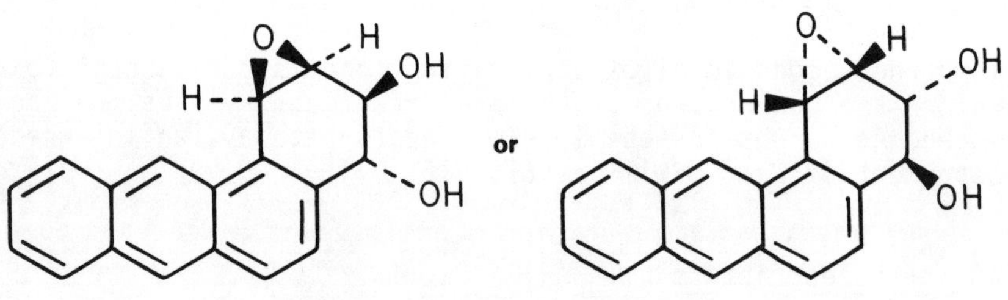

IUPAC: (1*R, 2*S**, 3*S**, 4*R*)-**
CAS: (1aα, 2β, 3α, 11cα)-
Other: *anti*-

FIGURE 16. Stereodescriptors for the relative configurations.

CAS Absolute Stereodescriptors

To describe the absolute configurations of systematically named compounds, the CAS system is based on analogy with the carbohydrate system for describing stereoisomers. A chiral reference center is defined and the remaining chiral centers are related to it by a relative descriptor. For example in the carbohydrate descriptor D-gluco, D describes the reference center and gluco describes the relative configurations of the chiral centers. Analogously, among systematic stereochemical descriptors the absolute terms R and S are employed for a single "reference" center, and the following relative terms are cited under various conditions: cis, trans, exo, endo, syn, anti, α, β, R*, S*, E, and Z.

The absolute configurations of the (+)- and (-)-syn, and the (+)- and (-)-anti isomers as described in the CAS system are compared in Figure 14. The reference center for the CAS system is the first cited chiral center, which is position "1a" of the ring, numbered according to the fusion name 1a,2,3,11c-tetrahydrobenzo[6,7]phenanthro[3,4-b]oxirene-2,3-diol. The relative configuration is described by α and β.

CONCLUSION

Each of the three nomenclature systems documented in the IUPAC rules for organic nomenclature (substitutive, additive, and fusion) can be used to describe the IUPAC metabolic products of PAH, such as the benz[a]anthracene metabolites. One must exercise care in forming these names so that the correct stoichiometry is expressed. Class names and abbreviations should also reflect correct nomenclature practices.

The relative stereodescriptors syn and anti have no official definition at present. For interdisciplinary communication, the IUPAC stereodescriptors R* and S* are used to describe relative configurations, while R and S are used to describe absolute configurations. The CAS system for describing stereoisomers is based on an analogy to carbohydrate stereodescriptors: a single reference center is described absolutely, which is then added to the complete relative descriptor.

REFERENCES

1. Loening, K. L. and Merritt, J. E. (1983): Some aids for naming polycyclic aromatic hydrocarbons and their heterocyclic analogs. In: Polynuclear Aromatic Hydrocarbons: Formation, Metabolism, and Measurement, edited by M. Cooke and A.J. Dennis, pp. 819-843, Battelle Press, Columbus, Ohio.
2. Jacob, J. and Loening, K. L. (1983): The nomenclature of PAH. Presented at the 8th International Symposium on Polynuclear Aromatic Hydrocarbons, Columbus, Ohio, 27 October 1983.
3. Later, D. W., Wright, C. W., Loening, K. L., and Merritt, J. E. (1986): Systematic nomenclature of the nitrogen, oxygen, and sulfur functional polycyclic aromatic compounds. In: Polynuclear Aromatic Hydrocarbons: Chemistry and Carcinogenesis, edited by M. Cooke and A. J. Dennis, Battelle Press, Columbus, Ohio.
4. Yang, S. K., Mushtaq, M. and Chiu, P. L. (1985): Stereoselective metabolism and activations of polycyclic aromatic hydrocarbons. In: Polycyclic Hydrocarbons and Carcinogenesis (ACS Symposium Series No. 283), edited by R. G. Harvey, p. 26, American Chemical Society, Washington, D.C.
5. IUPAC (1979): Nomenclature of organic chemistry, Sections A,B,C,D, E, F and H. Pergamon, Oxford. 559 pp. (The Blue Book)
6. IUPAC (1971): Nomenclature of inorganic chemistry. 2nd ed., Butterworths, London. 110 pp. (The Red Book)
7. IUPAC (1978): Compendium of analytical nomenclature. Pergamon, Oxford, 223 pp. (The Orange Book)
8. IUPAC (1979): Manual of symbols and terminology for physicochemical quantities and units. Revised ed. Pergamon, Oxford. 41 pp. (The Green Book)
9. IUPAC-IUB (1978): Biochemical nomenclature and related documents. The Biochemical Society, London. 223 pp. (The Compendium).
10. IUB (1984): Enzyme nomenclature 1984. Academic Press, Orlando, Florida. 646 pp.
11. Gillilan, R. E. (1982): Substituent effects on rates and product distributions in the hydrolysis reactions of naphthalene tetrahydro epoxides. Models for tetrahydro epoxides of polycyclic aromatic hydrocarbons, J. Am. Chem. Soc., 104:4481-4482.
12. LeBreton, P. R. (1985): The intercalation of benzo[a]pyrene and 7,12- dimethylbenz[a]anthracene metabolites and metabolite model compounds into DNA. In:

Polycyclic Hydrocarbons and Carcinogenesis (ACS Symposium Series No. 283), edited by R. G. Harvey, pp. 213, American Chemical Society, Washington, D.C.

13. Cooke, M. and Dennis, A. J. (1983): Key word index. In: Polynuclear Aromatic Hydrocarbons: Formation, Metabolism, and Measurement, Battelle Press, Columbus, Ohio.

14. Beland, F. A. and Harvey, R.G. (1976): The isomeric 9,10-oxides of trans-7,8-dihydroxy-7,8-dihydrobenzo[a]pyrene. J. Chem. Soc, Chem. Commun., 84-85.

15. Chemical Abstracts Index Guide (1985): Appendix IV, Section 203.

16. Cahn, R. S., Ingold, C. and Prelog, V. (1966): Specification of molecular chirality. Angew. Chem., Int. Ed. Engl. 5(4), 385-415.

17. Yagi, H., Akagi, H., Thakker, D. R., Mah, H. D., Koreeda, M. and Jerina, D. M. (1977): Absolute stereochemistry of the highly mutagenic 7,8-diol 9,10-epoxides derived from the potent carcinogen trans-7,8-dihydroxy-7,8-dihydrobenzo[a]pyrene. J. Am. Chem. Soc, 99(7) 2358.

Chapter 4

ENZYME NOMENCLATURE

JOY E. MERRITT AND KURT L. LOENING

THE IUPAC-IUB ENZYME NOMENCLATURE BOOK

The concept of an enzyme and the discovery of enzymes date back to the 19th century. The word "enzyme" was introduced in 1898, but the purification of enzymes was not accomplished until 1920 (1,2). Today there are over 2477 enzymes listed in the most recent compilation by the International Union of Biochemistry (IUB) (3). This contrasts with only 712 enzymes identified in the first report of the International Enzyme Commission in 1961 (4).

The International Enzyme Commission was established in 1956 by IUB in consultation with the International Union of Pure and Applied Chemistry (IUPAC). Its task was to bring order to the naming of enzymes. Names coined by individuals had resulted in many cases where the same enzyme became known by several different names, and similar names were sometimes given to enzymes of quite different types. Many of the names gave no indication of the nature of the reactions catalyzed, and some of the names in use were misleading. The Commission was dissolved after publication of their report in 1961(4). Eventually the function of the Commission passed on to the IUB Nomenclature Committee which today keeps track of reports of new enzymes, determines their classification, and provides recommended names.

The latest edition of the IUB/NC Enzyme Nomenclature was published in 1984, and two supplements have been published (5). The first report was titled simply, Report of the Commission on Enzymes (4). However with subsequent editions (6,7,8), the title was changed to Enzyme Nomenclature, and a subtitle added which has evolved from "...Classification of Enzymes" to the current subtitle, "Classification of Enzyme-Catalysed Reactions".

Enzyme Nomenclature Based on Activity, Not Structure

An enzyme is a protein with catalytic properties due to its power of specific activity. Structurally, enzymes are proteins. Thus, for an enzyme fully characterized by structure, a very systematic, lengthy chemical name based on its amino acid sequence can be generated in accord with the IUPAC-IUB rules for amino acids and peptides (Figure 1)(9). However, outside of the capabilities of a machine searchable database, these names are not very practical. Thus, for purposes of communication in the scientific literature shorter names are preferred.

The detection of an enzyme through observation of catalysis often precedes the complete structural elucidation of the enzyme structure involved. Also, an enzyme may not be a single protein entity, but a family of entities, all closely related, all catalyzing the same reaction. A specific catalytic activity may be discovered in two different sources, for example in a plant and in an animal, but the isolated proteins responsible for the catalytic activity may differ structurally. What the proteins may have in common is a similarity in active sites.

Thus, the Commission decided to develop a classification and nomenclature based on similarities of the reactions catalyzed rather than on the cumbersome amino-acid sequence names.

As a result, enzyme names and numbers refer to biological or catalytic activities and not to structures.

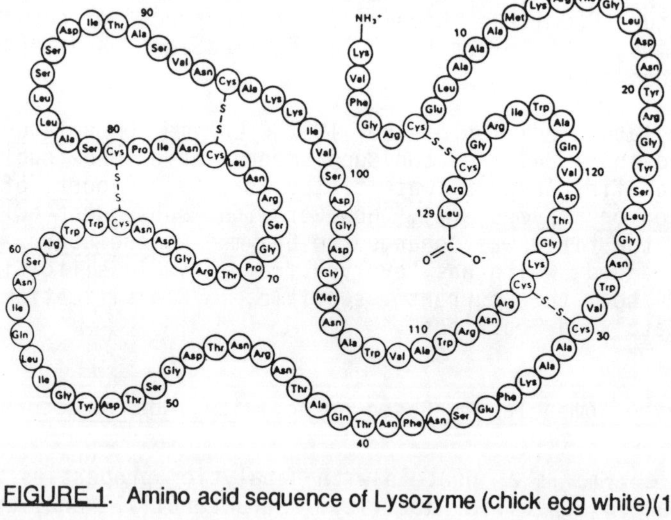

FIGURE 1. Amino acid sequence of Lysozyme (chick egg white)(10).

ENZYME NOMENCLATURE

Number	Recommended Name	Reaction
3.5.1.6	β-Ureidopropionase	*N*-Carbamoyl-β-alanine + H_2O = β-alanine + CO_2 + NH_3
3.5.1.7	Ureidosuccinase	*N*-Carbamoyl-L-aspartate + H_2O = L-aspartate + CO_2 + NH_3
3.5.1.8	Formylaspartate deformylase	*N*-Formyl-L-aspartate + H_2O = formate + L-aspartate
3.5.1.9	Arylformamidase	*N*-Formyl-L-kynurenine + H_2O = formate + L-kynurenine

369

Number	Other Names	Basis for classification (Systematic Name)	Comments	References
3.5.1.6		*N*-Carbamoyl-β-alanine amidohydrolase	The animal enzyme also acts on β-ureidoisobutyrate	515, 530
3.5.1.7		*N*-Carbamoyl-L-aspartate amidohydrolase		2243
3.5.1.8		*N*-Formyl-L-aspartate amidohydrolase		2844
3.5.1.9	Kynurenine form-amidase, Formylase, Formylkynureninase	Aryl-formylamine amidohydrolase	Also acts on other aromatic formylamines	1441, 1719, 2497

FIGURE 2. Excerpt of two facing pages in Enzyme Nomenclature.

CONTENTS

The list of almost 2500 enzymes comprises the bulk of the book. For each enzyme listed specific information with regard to its nomenclature and classification is given in seven columns stretched across two facing pages (Figure 2). These columns are labeled: 1. Number (Enzyme Commission (E.C.) Numbers), 2. Recommended Name, 3. Reaction, 4. Other Names, 5. Basis for classification (Systematic Name), 6. Comments (about its source and range of specificity), and 7. References.

Column 1 - Enzyme Commission Number

The E.C. Numbers are code numbers which contain four elements separated by points. For example, 1.10.1.1 is the E.C. Number for trans-Acenaphthene-1,2-diol dehydrogenase. The first number shows to which of the six main divisions the enzyme belongs, in this case oxidoreductases (Figure 3). Each of these classes is defined in more detail later. The second and third elements of the E.C. Number indicate subdivision and sub-subdivisions unique to each of the main classes. These are illustrated by an excerpt from the section on transferases (Figure 4). Finally, a fourth number is added to the sub-subclassification number. It is merely a serial number unique to the individual enzyme.

1. Oxidoreductases	4. Lyases
2. Transferases	5. Isomerases
3. Hydrolases	6. Ligases

FIGURE 3. The six main classes of enzymes.

Columns 2 and 4 - Recommended and Other Names

Recommended names are trivial names. Often for new enzymes they are shortened versions of the systematic name.

2. **Transferases**

   2.1 Transferring one-carbon groups
      2.1.1 Methyltransferases
      2.1.2 Hydroxymethyl-, formyl-, and related transferases
      2.1.3 Carboxyl- and carbamoyltransferases
      2.1.4 Amidinotransferases

   2.2 Transferring aldehyde or ketone residues

   2.3 Acyltransferases
      2.3.1 Acyltransferases
      2.3.2 Aminoacyltransferases

   2.4 Glycosyltransferases
      2.4.1 Hexosyltransferases
      2.4.2 Pentosyltransferases
      2.4.99 Transferring other glycosyl groups

   2.5 Transferring alkyl or aryl groups, other than methyl groups

   2.6 Transferring nitrogenous groups
      2.6.1 Aminotransferases
      2.6.3 Oximinotransferases
      2.6.99 Transferring other nitrogenous groups

   2.7 Transferring phosphorus-containing groups
      2.7.1 Phosphotransferases with an alcohol group as acceptor
      2.7.2 Phosphotransferases with a carboxyl group as acceptor
      2.7.3 Phosphotransferases with a nitrogenous group as acceptor
      2.7.4 Phosphotransferases with a phosphate group as acceptor
      2.7.6 Diphosphotransferases
      2.7.7 Nucleotidyltransferases
      2.7.8 Transferases for other substituted phosphate groups
      2.7.9 Phosphotransferases with paired acceptors

   2.8 Transferring sulfur-containing groups
      2.8.1 Sulfurtransferases
      2.8.2 Sulfotransferases
      2.8.3 CoA-transferases

3. **Hydrolases**

   3.1 Acting on ester bonds
      3.1.1 Carboxylic ester hydrolases
      3.1.2 Thiolester hydrolases
      3.1.3 Phosphoric monoester hydrolases

FIGURE 4. Subclassification of the Transferases.

But in most instances they are the trivial names selected and approved by the Commission for common use. Trivial names other than those recommended by the Commission are listed in the book in the column labeled "Other Names", and their use is discouraged. For example, Catechol oxidase is a recommended name. Its longer systematic name is 1,2-Benzenediol:oxygen oxidoreductase (E.C. 1.10.3.1). Other names that are acknowledged but not recommended are Diphenol oxidase, *o*-Diphenolase, Phenolase, Polyphenol oxidase, Tyrosinase.

Column 3 - Reaction

The reaction catalyzed is the basis for both the classification and systematic nomenclature of enzymes. Because most enzymes can theoretically catalyze a reaction in either direction depending on the reaction conditions, the equal sign is used in place of the usual reaction arrows. The reaction equations as recorded are not intended to imply the direction of the reaction. The reaction catalyzed by Catechol oxidase is recorded as

2 Catechol + O_2 = 2 1,2-benzoquinone + 2 H_2O

Column 5 - Systematic Name

Actually systematic names are quite logical and the rules for their formation within each of the main classes are quite simple. Usually they consist of two parts. The first part refers to the substrate(s), the second part refers to the action involved. The specific rules for their formation are given in the front of the book. They are based on the classification of enzyme reaction types discussed and illustrated with specific examples below.

Column 6 - Comments

The comments usually refer to the specificity of the enzyme toward its substrates.

Column 7 - References

The references are a very select list. They are intended to provide sufficient evidence for the existence of an enzyme catalyzing the reaction as set out. They do not necessarily document the first description of the enzyme. Major reviews are preferred. There are more than one reference in some instances because they document the enzyme in animal, plant and bacterial systems.

ENZYME REACTION TYPES

1. Oxidoreductases

To the oxidoreductase class belong all enzymes catalyzing oxido-reductions, and the substrate oxidized is regarded as hydrogen or electron donor. The recommended format for the systematic name is donor:acceptor oxidoreductase (Figure 5).

Alcohol dehydrogenase illustrates this class. This enzyme is mainly found in the liver. Its primary substrate is ethanol. It is the enzyme responsible for eliminating alcohol from the body. The aldehydes and ketones produced by the oxidation of ethanol contribute to hangovers. In the other direction this enzyme catalyzes the last step in alcohol fermentation by yeast.

Thus, for the reaction illustrated the alcohol is the donor, and NAD is the acceptor so the systematic name is alcohol:NAD oxidoreductase.

It should be mentioned that NAD is a cofactor (or coenzyme). The reaction does not take place efficiently without it. Because it is so common in this class of reactions, its presence is understood, so that it may be dropped from the systematic name.

As stated above, one of the general principles adopted by the Commission is that it makes no difference which way the reaction actually occurs, i.e. whether it is an oxidation or a reduction reaction that takes place, it is the overall type of reaction that determines its place in the classification.

Format for Systematic Name:

donor:acceptor oxidoreductase

Recommended Names:

dehydrogenase

acceptor reductase

oxidase (only in cases where O_2 is an acceptor)

Example: EC 1.1.1.1

Catalyzed Reaction:

$$NAD^+ + R\text{-}\underset{\underset{OH}{|}}{\overset{\overset{H}{|}}{C}}\text{-}R' \rightleftharpoons NADH + R\text{-}\underset{\underset{O}{\|}}{C}\text{-}R' + H^+$$

NAD^+ is the oxidized form of nicotinamide adenine dinucleotide.
NADH is the reduced form of nicotinamide adenine dinucleotide.

Systematic Name: Alcohol:NAD^+ oxidoreductase

Recommended Name: Alcohol dehydrogenase

Classification:

1. Oxidoreductase
1.1 Acting on the CH-OH group of donors
1.1.1 with NAD^+ or $NADP^+$ as acceptor

$NADP^+$ is the oxidized form of nicotinamide adenine dinucleotide phosphate

FIGURE 5. Class 1. The Oxidoreductases.

2. Transferases

Transferases are enzymes transferring a group such as a methyl group or a glycosyl group, from one compound (generally regarded as donor) to another compound (generally regarded as acceptor) (Figure 6). The format for the systematic name is "donor:acceptor grouptransferase". For example, the enzyme that transfers a methyl group from S-adenosylmethionine to nicotinamide is named systematically S-Adenosyl-L-methionine: nicotinamide N-methyltransferase.

Format for Systematic Name:

donor:acceptor grouptransferase

Recommended Names:

acceptor grouptransferase

donor grouptransferase

Example:

Catalyzed Reaction:

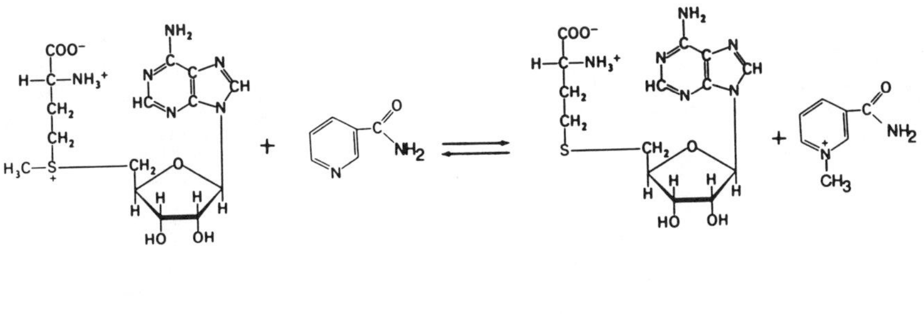

S-Adenosylmethionine + nicotinamide ⟶ S-adenosyl-L-homocysteine + 1-methylnicotinamide

Systematic Name: S-Adenosyl-L-methionine:nicotinamide N-methyltransferase

Recommended Name: Nicotinamide methyltransferase

Classification: EC 2.1.1.1
2. Transferases
2.1. Transferring one-carbon groups
2.1.1. Methyltransferases

FIGURE 6. Class 2. The Transferases.

3. Hydrolases

Hydrolases catalyze the hydrolysis of various bonds (Figure 7). They break larger molecules into smaller ones by the overall process of hydrolysis. The format of the systematic name is "substrate grouphydrolase". Thus, the systematic name of the enzyme that catalyzes the hydrolysis of urea into carbon dioxide and ammonia is called urea amidohydrolase. The recommended name is simply the name of the substrate with the suffix "-ase". Urease is found primarily in plants, as it takes part in the urea cycle.

One of the more important groups of enzymes in this class includes the proteolytic enzymes such as trypsin, chymotrypsin, and pepsin which aid in the digestion of starches, fats and proteins.

Format for Systematic Name:

substrate grouphydrolase

Recommended Names:

name of the substrate with the suffix -ase

Example:

Catalyzed Reaction:

$$H_2N\text{-}\overset{\overset{\large O}{\|}}{C}\text{-}NH_2 \longrightarrow CO_2 + NH_3$$

Urea — Carbon dioxide — Ammonia

Systematic Name: Urea amidohydrolase

Recommended Name: Urease

Classification: EC 3.5.1.5
3. Hydrolases
3.5. Acting on Carbon-Nitorgen Bonds, Other Than Peptide Bonds
3.5.1. In Linear Amides

FIGURE 7. Class 3. The Hydrolases.

4. Lyases

Lyases are cleaving enzymes. They cleave C-C, C-O, and C-N bonds by means other than hydrolysis or oxidation, perhaps by elimination, creating double bonds, or perhaps by addition by adding groups to double bonds. They differ from other enzymes in that two substrates are involved in one reaction direction, but only one in the other direction. The systematic name is patterned after the format substrate group-lyase (Figure 8). The Commission recommends that the hyphen be retained in the name.

In the reaction illustrated, a carboxy group is cleaved from acetoacetate so its systematic name is acetoacetate carboxy-lyase. Its recommended name is acetoacetate decarboxylase.

Format for Systematic Name:

substrate group-lyase

Recommended Names:

decarboxylase

aldolase

dehydratase

synthase

Example:

Catalyzed Reaction:

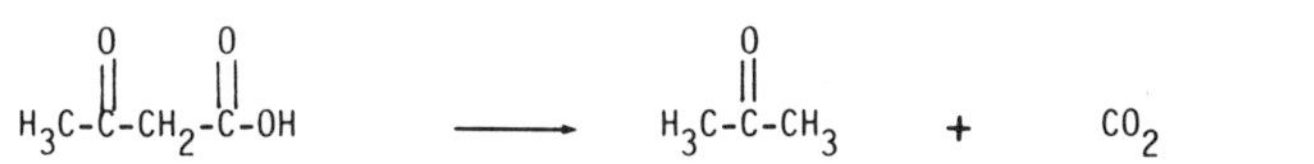

Acetoacetate — Acetone — Carbon dioxide

Systematic Name: Acetoacetate carboxy-lyase

Recommended Name: Acetoacetate decarboxylase

Classification: EC 4.1.1.4
4. Lyases
4.1. Carbon-Carbon Lyases
4.1.1. Carboxy-Lyases

FIGURE 8. Class 4. The Lyases.

5. Isomerases

Isomerases catalyze geometric or structural changes within one molecule (Figure 9). According to the type of isomerism involved they are named by combining the name of the substrate with a term such as isomerase, racemase, epimerase, mutase, etc. Thus in converting a cis double bond to a trans double bond the enzyme is called maleate *cis*-*trans*-isomerase. The -ate ending is favored by the Commission for a number of reasons. One is that it is more easily translated into some languages than is -ic acid. Also it is believed that the physiological conditions favor the anionic form.

Format for Systematic Name:

substrate isomerase

substrate racemase

substrate epimerase

substrate mutase

Recommended Names:

In general, similar to the systematic names.

Catalyzed Reaction:

Maleate Fumarate

Systematic Name: Maleate *cis*-*trans*-isomerase

Recommended Name: Maleate isomerase

Classification: EC 5.2.1.1
5. Isomerases
5.2. *cis*-*trans*-Isomerases
5.2.1. (not further divided at present)

FIGURE 9. Class 5. The Isomerases.

6. Ligases

Ligases are anabolic enzymes. They catalyze the joining together of two molecules with concomitant hydrolysis of the pyrophosphate bond in ATP or a similar triphosphate (Figure 10). The format for the systematic name is X:Y ligase (ADP-forming). In the example selected, the enzyme catalyzes the transformation of glutamate in the presence of ammonia into glutamine. To do this requires the energy expended in the hydrolysis of the pyrophosphate bond in ATP to ADP. Thus, this enzyme is called systematically glutamate:ammonia ligase.

Systematic Names:

X:Y ligase (ADP-forming).

Recommended Names:

X - Y ligase

synthase

Catalyzed Reaction:

ATP + L-glutamate + NH_3 ⟶ ADP + orthophosphate + L-glutamine

Systematic Name: L-Glutamate:ammonia ligase (ADP-forming)

Recommended Name: Glutamate-ammonia ligase

Other Name: Glutamine synthetase

Classification: EC 6.3.1.2
- 6. Ligases
- 6.3. Forming Carbon-Nitrogen Bonds
- 6.3.1. Acid-Ammonia (or Amine) Ligases (Amide Synthases)

FIGURE 10. Class 6. The Ligases.

EDITORIAL PRACTICES

The use of the systematic name has certain advantages over the use of EC Numbers. Code numbers are only interpreted with the aid of the Enzyme book, whereas systematic names are self-explanatory. Systematic names can be formed by the discoverer, but code numbers should not be assigned by individuals.

The Commission recommends that where an enzyme is the main subject of a paper or abstract, its code number (*i.e.*, E.C. Number), systematic name, or alternatively, the reaction equation and source should be given at its first mention; thereafter the recommended name should be used. One journal that has adopted this practice is the *European Journal of Biochemistry*. In each issue, at the bottom of the first page of each article involving one or more enzymes, a list of EC Numbers and corresponding names is published.

ADDITIONS, DELETIONS AND CHANGES OF CLASSIFICATION

IUB depends on the biochemical public for suggestions for additions, deletions, and changes in classifications. A form is available for proposing new entries in the back of the book, which is reproduced here for your general information (Figure 11). Chemical Abstracts (CA) is the biggest contributor for suggestions or additions to the enzyme list. Each year CA compiles a list of new enzymes encountered in the literature to send to IUB. This year a list of about 300 new ones has been compiled.

HISTORY

The editorial work is carried out in the office of Edwin C. Webb. He was secretary of the IUB Standing Committee on Enzymes from 1961-1968. Both at the University of Queensland and at the Macquarie University in Australia, he has overseen entries into the list for 26 years. He is now retired, and still continues the collection and evaluation of proposed new

REPORT FORM ON AN ENZYME NOT INCLUDED IN THIS EDITION OF ENZYME NOMENCLATURE

The Nomenclature Committee of the International Union of Biochemistry intends to update the Enzyme List from time to time by the publication of Supplements, and ultimately by the production of a full new edition. The assistance of the biochemical community is sought in this task. This tear-off sheet can be used to draw the attention of the editor to enzymes missing from this list, or to errors in existing entries.

THIS SHEET IS NOT COPYRIGHT. Photocopies can be used for additional contributions.

Reaction catalysed:

Name proposed by authors (*if any*):

Subclass in Enzyme Nomenclature proposed (*if any*) (*e.g.* 2.7.7.):

Source of enzyme: (*e.g.* yeast, horse liver, *E. coli, etc.*)

Brief comment on specificity:

Cofactor requirement:

Reference: (*If accepted by a journal but not yet published, give name of journal and date of acceptance*)

Name and address of person submitting this report:

The completed form should be sent to: Professor Edwin C. Webb
1/221 King's Road
Mundingburra
Townsville, Queensland 4812
Australia

FIGURE 11. IUB report form for an enzyme.

enzymes to add to the list. He is also known for his text coauthored with Malcolm Dixon titled simply, Enzymes first published in 1958 and which is now in its third edition (1,2). Malcolm Dixon was president of the Enzyme Commission from 1956-1961. Included in the first edition is a 43 page comprehensive table of enzymes that were known at the time. In the third edition, this table now takes up 190 pages.

Prior to the publication of Webb's table of enzymes, O. Hoffmann-Ostenhof published a review in 1951 on the problem of enzyme classification and nomenclature with specific proposals(11). O. Hoffmann-Ostenhof was Secretary of the Enzyme Commission, and chairman of the Commission on Biochemical Nomenclature from 1965 to 1976 (12). It is to the credit of the International Commission on Enzymes that the efforts of these men were brought together resulting in the first international recommendations on enzyme nomenclature and classification.

REFERENCES

1. Dixon, Malcolm and Webb, Edwin C. (1958): Enzymes. Academic Press, New York, 782 pp.

2. Dixon, Malcolm, Webb, Edwin C., Thorne, C.J.R., and Tipton, K.F. (1979): Enzymes. 3rd ed. Academic Press, New York, 1116 pp.

3. International Union of Biochemistry. Nomenclature Committee. (1984): Enzyme Nomenclature. Recommendations 1984 on the Nomenclature and Classification of Enzyme-Catalysed Reactions. Academic Press, Orlando, Florida, 1984. 646 pp.

4. International Union of Biochemistry. (1961): Report of the Commission on Enzymes. Pergamon Press, New York, 159 pp.

5. International Union of Biochemistry. Nomenclature Committee. "Enzyme Nomenclature". Recommendations 1984. Supplement 1 (1986): Corrections and Additions. Eur. J. Biochem. 157: 1-26. Supplement 2 (1989): Corrections and Additions. Eur. J. Biochem. 179: 489-533.

6. International Union of Biochemistry. (1965): Enzyme Nomenclature. Recommendations 1964 on the Nomenclature and Classification of Enzymes together with their Units and the the Symbols of Enzyme Kinetics. Elsevier, Amsterdam, 219 pp.

7. International Union of Pure and Applied Chemistry, and the International Union of Biochemistry. Commission on Biochemical Nomenclature. (1973): Enzyme Nomenclature. Recommendations 1972 on the Nomenclature and Classification of Enzymes together with their Units and the Symbols of Enzyme Kinetics. Elsevier, Amsterdam, 443 pp.

8. International Union of Biochemistry. Nomenclature Committee. (1979): Enzyme Nomenclature. Recommendations 1978 on the Nomenclature and Classification of Enzymes. Academic Press, New York, 606 pp.

9. International Union of Pure and Applied Chemistry, and the International Union of Biochemistry. Joint Commission on Biochemical Nomenclature. (1984): "Nomenclature and Symbolism for Amino Acids and Peptides". Recommendations 1983. Eur. J. Biochem. 1984, 128, 9-37. Corrections (1985): Eur. J. Biochem. 152, 1.

10. Stryer, Lubert (1975): Biochemistry. W. H. Freeman and Company, San Francisco, p. 137.

11. Hoffmann-Ostenhof, O., (1952): "Suggestions for a More Rational Classification and Nomenclature of Enzymes". Adv. Enzymol. 14, 219-260.

12. Karlson, P., (1984): "Otto Hoffmann-Ostenhof zum 70. Geburtstag". Naturwissenschaften 71, 491-492.